Practical
Air Conditioning
Equipment Repair

Practical
Air Conditioning
Equipment Repair

Anthony J. Caristi

McGraw-Hill Book Company

New York St. Louis San Francisco Auckland Bogotá
Hamburg London Madrid Mexico Milan Montreal
New Delhi Panama Paris São Paolo
Singapore Sydney Tokyo Toronto

Library of Congress catalog card number 90-84559

10 9 8 7 6 5 4 3 2 1

ISBN 0-07-009929-4

Intertext Publications / Multiscience Press, Inc.
One Lincoln Plaza
New York, NY 10023

McGraw-Hill Book Company
1221 Avenue of the Americas
New York, NY 10020

Composed in PageMaker by Horizon Publications, San Diego, CA

To my wife, Betty, who has graciously tolerated the long hours spent in the preparation of this manuscript.

Contents

Preface

The advance of high technology has brought about a standard of living which few people could imagine even just a couple of decades ago. All kinds of inventions, devices, and appliances which were once considered luxuries are now commonplace, and have become necessities. Perhaps one of the best examples of this is air conditioning, which is now standard equipment in nearly every commercial or business establishment. Even our homes and automobiles are considered incomplete unless some form of cooling is installed. Many of us can remember our childhood, when air conditioning was rarely found. Today, we have become so accustomed to climate control all year round that life as we know it would be impossible without some form of cooling during the summer months.

As air conditioning equipment has become more sophisticated, so has the need for more and better-trained mechanics and technicians to keep the complex equipment in running order. But the supply of well-trained air conditioning repair personnel has barely kept up with the demand, and the situation is sure to get more critical as time goes on.

This book has been written with this situation in mind. It is hoped that it can provide some insight into the workings of air conditioning systems so that the reader, with or without formal training, can perform some of the repair techniques which can place a malfunctioning air conditioning system back in service. Emphasis has been placed on the practical aspect of troubleshooting and repair so that the system can be placed back in running condition in short time, not only by the professional, but also by those who do not have formal air conditioning training or experience. Most breakdowns seem to occur at the worst time, such as during a heat wave when air conditioning service personnel are hard to get. It's the kind of a situation where the job can't be put off for a week or two.

Much effort has been taken to write this book in easy-to-read language, avoiding complex theory and mathematics. It can easily be understood by anyone who is familiar with repair procedures and is handy with ordinary tools. The air conditioning principles detailed in this book do not have to be fully understood for the layman to successfully repair or service an air conditioning unit.

Chapter 1 starts with air conditioning principles and describes the various components of an air conditioning unit or system. This is followed in Chapter 2 with the theory of air conditioning heat loads. Sufficient information is presented here to enable the reader to calculate the proper size unit for any given application, without the use of complex mathematical expressions.

Chapter 3 explains in detail the workings of each of the various components that may be encountered when servicing an air conditioner. Included are detailed discussions of residential and commercial air conditioning systems, heat pumps, and ammonia absorption systems.

Air conditioning service and repair equipment is covered in Chapter 4, as well as detailed information on troubleshooting A/C units and systems. Chapter 5 includes a thorough discussion on service and repair procedures for the compression cycle air conditioner.

Subsequent chapters cover the service and repair of evaporative coolers, dehumidifiers, heat pumps, and ammonia absorption systems. Each of these subjects is discussed in detail. Much emphasis has been placed on the description and theory of the various components which make up these systems.

Special attention has been given to the automotive air conditioning system. Although vehicle air conditioning systems operate on the same principles as residential and industrial units, they have their own special problems and repair techniques. This material is presented in detail.

Five appendices and a glossary provide valuable information which can be very helpful, not only in understanding air conditioning concepts, but also as a handy source of reference material.

Although the information presented in this air conditioning troubleshooting and repair manual can be found in many different publications, it is hoped that this will be the one book that will be referred to for quick and easy information to solve that air conditioning problem.

The information presented in this book has been gathered from many sources, including many companies that manufacture air conditioning equipment, tools, and supplies. The author wishes to thank York International Corporation, The Dometic Corporation, Tecumseh Products Company, Lennox Industries, Inc., Lucas-Milhaupt, Inc., and all the other fine companies which helped make this book possible.

The author wishes to give special thanks to T. J. Byers, who provided much help in the production of this book, including the chapter on evaporative cooling.

Anthony J. Caristi

Practical
Air Conditioning
Equipment Repair

Principles of Air Conditioning

1.1 General Information

The term air conditioning usually refers to an electromechanical system in which the ambient temperature and humidity is controlled and reduced to a level which is comfortable for human occupancy, or in some cases, to provide improved operating conditions for sensitive electrical equipment such as computers. Some systems are designed as combination heating and cooling, and provide year-round temperature control. The method by which cooling is accomplished is usually through refrigeration equipment, which has been specially designed to provide an ambient temperature differential of about 15 to 20 degrees F (about 10 degrees C) between the inside and outside of a controlled area.

Heat pumps are devices which are used to provide summer cooling and winter heating by transferring heat energy from a cool environment to a warmer one. Even at freezing temperatures, the air contains sufficient heat energy which can be captured and moved to the inside of a building. Specialized controls within the unit allow the direction of heat transfer to occur in either direction in accordance with the season. These systems are very similar to compression cycle air conditioning units and are becoming a viable alternative to other forms of heating systems, especially in those areas where winter temperatures do not go much below freezing.

Before electromechanical air conditioning systems using the principles of refrigeration became economically feasible, cooling of commercial establishments, such as movie theaters, was accomplished by forcing large volumes of air across a bed of ice. The resulting cooling of the air caused by the melting of the ice provided relief from the hot, humid summer temperatures. This is where the term "air cooled," which was prominently displayed on theater marquees, became well known.

As illustrated in Figure 1.1, the use of ice to provide air cooling was very successful, and the amount of cooling could be controlled by using more or less air flow or by melting greater or smaller quantities of ice. This system was effective in humid atmospheres since much of the water content of the air would condense on the ice.

In areas of the country where humidity levels are naturally low during the summer, such as in the southwest, another form of air conditioning is employed. This is the evaporative cooler, which operates on the principle of an evaporating liquid (water in this case) which removes heat from the surrounding air. The principle is the same as if a small amount of rubbing alcohol were placed on to the skin. The cooling effect due to the evaporation of the alcohol would be immediately felt.

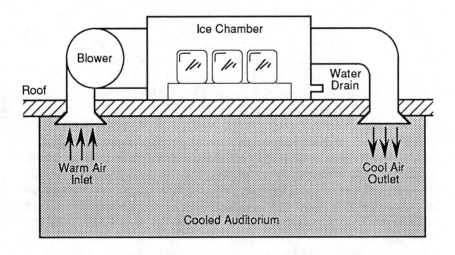

Figure 1.1 Air Cooling Provided by Melting Ice

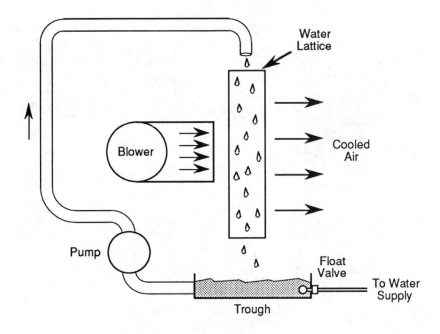

Figure 1.2 Evaporative Cooling System

Evaporative cooling systems (Figure 1.2) will work well only in locations of low humidity since they add moisture to the air as the cooling effect takes place. This can actually be an advantage, since the humidity level in some areas can be as low as 5% during the hot summer weather. Locations of high humidity, such as the eastern part of the country, cannot be satisfactorily cooled by the evaporative process.

Mechanical household refrigeration systems for preservation of food first appeared in the early part of the 20th century. It wasn't until the late 1920s that such systems were designed to be used for summer cooling. The first systems that were designed employed compressors which were belt-driven by external motors. By 1940, the hermetically sealed compressor made its way into the rapidly expanding air conditioning industry. A typical mechanical refrigeration system is illustrated in Figure 1.3.

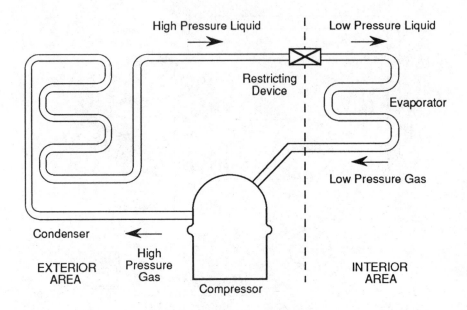

Figure 1.3 Mechanical Refrigeration Cycle

The absorption system of refrigeration has been used for many decades as an alternative to mechanical-compression-type refrigerators. This system is also capable of being used for producing very large cooling capacities for residential and commercial air conditioning needs, and is widely used. The absorption system uses heat instead of mechanical energy to provide the refrigeration process. The refrigerant is usually ammonia and the absorbent is water.

1.2 CFC Refrigerants and the Environment

Common refrigerants such as R-12 and R-22 have been used for many years in residential, commercial, and automotive air conditioning systems, and have proven to be safe and low in cost. However, these chlorofluorocarbons (CFCs) are implicated in the depletion of the earth's ozone layer and environmental regulations have been passed to totally phase out the use of certain refrigerants by the year 2000.

In order to reduce the demand for certain ozone-depleting refrigerants such as R-12, the federal government has imposed a tax of $1.37 per pound on that product starting in the year 1990. The tax burden will be increased in following years to reduce demand for R-12 and facilitate the use of alternative refrigerants as they become available. It is foreseeable that new taxes and restrictions will be placed on all environmentally damaging products being used today.

Companies such as Du Pont (E.I. du Pont de Nemours & Company, Inc.) plan to commercialize a series of alternative refrigerants, known as hydrochlorofluorocarbons (HCFCs) and hydrofluorocarbons (HFCs). These products are "friendly" to the environment since they exhibit less ozone-depletion potential and global warming characteristics than the refrigerants they are designed to replace.

Possible interim replacement products for R-12 include HFC-134A and ternary blends which are composed of HCFC-22 plus two other compounds. These are currently identified as KCD-9430 and KCD-9433. The latter shows a dramatic environmental improvement over R-12 with an ozone-depletion potential of only 3% (R-12 is standardized at 100%).

An alternative that is being considered for R-11 is a product known as HCFC-123. This product, and others which may be developed in the future, is similar to but not exactly the same as R-11, which it is designed to replace.

The service technician must be aware that substitute refrigerants can not be "dropped into" a system designed for a CFC refrigerant. Differences in properties may cause inefficient operation or even failure of the equipment. Any replacement refrigerant must be compatible with the oil in the system, as well as with any possible elastomeric or plastic components. Additionally, the sizing of the expansion or restricting device must be considered.

Before retrofitting any system with a substitute refrigerant, the manufacturer should be consulted to determine what changes must be made to ensure proper equipment performance.

1.3 Fundamentals of BTUs

BTU is an abbreviation for British Thermal Unit, which is commonly used as a measure of thermal energy in air conditioning systems. One BTU represents the heat required to raise one pound of water, at its maximum density at 39 degrees F (4 degrees C), 1 degree F. An equivalent measure is the calorie, which is defined as the quantity of heat required to raise one gram of water 1 degree C. The calorie is an extremely small measure of heat energy, being equal to just 0.004 BTU. For this reason, it is generally not used in specifying air conditioning capacities.

When heat energy is used to raise the temperature of a substance, the effect is something which can be felt or measured. This is referred to as "sensible heat." However, it also takes heat energy to change the state of a substance from solid to liquid, and from liquid to gas.

This heat energy is called "latent" or hidden heat. Latent heat is defined as the heat energy which brings about a change of state without a change in temperature. Note that the concept of latent heat works in two directions. It takes 144 BTU of heat per pound of ice to melt it to water, and 144 BTU per pound of water must be extracted to change it to ice.

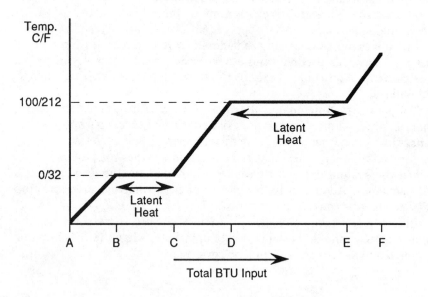

**Figure 1.4 Illustration of Heat Energy Requirement to
Change Ice into Superheated Vapor**

As illustrated in Figure 1.4, the cumulative heat input to the system is broken down into five stages. The temperature of the ice is brought up to the freezing level by the addition of sensible heat as represented by point A to point B. At the freezing temperature, 32 degrees F (0 degrees C), latent heat (B to C) is added to the system to melt the ice and change it to water, with no increase in temperature.

The temperature of the water is raised further, to the boiling point, by the addition of sensible heat from point C to point D. Then, the addition of latent heat from D to E causes the water to change to vapor, with no increase in temperature. Finally, sensible heat added between points E and F raises the temperature of the water vapor, and it becomes superheated.

A good example of latent heat is in the air cooling system described earlier, where air is forced over a large quantity of ice to provide cooling. In this case, the bulk of heat which is extracted from the air comes from the physical change of ice melting to water. This change in state from solid to liquid provides a known amount of BTUs of cooling (144 BTU per pound of ice).

It takes 288,000 BTU of latent heat energy to change one ton (2000 pounds) of ice at 32 degrees F to water at 32 degrees F. Note that during this change of state the temperature of the ice and water remains at 32 degrees F. The relationship of one ton of ice melting to provide 288,000 BTU of cooling is the basis for rating an air conditioner of 12,000 BTU as a 1 ton unit, since air conditioning systems are rated in BTU per hour. Thus, a 12,000 BTU unit, 1 ton rating, has the capacity to provide the equivalent cooling of 1 ton of ice changing to water in a 24-hour period.

Air conditioner efficiency is measured by a quantity called energy efficiency ratio (EER). This quantity is defined as the BTU-per-hour rating of the air conditioning unit divided by its power input in watts. This rating is now required by law to be indicated on every unit sold for household use in the United States. EER ratings for domestic air conditioning systems can be 10 or more. The higher the EER, the less power demand the unit will require for a given BTU size unit. Some low-efficiency units, with relatively poor EER ratings, cannot be legally sold in certain states. SEER, Seasonal Energy Efficiency Rating, is a similar measure of air conditioner efficiency and often referred to in place of EER.

1.4 Electrical Fundamentals

The most predominant electrical energy used worldwide is alternating current (AC), which is easily transmitted over long distances and can be transformed into low or high voltage by highly efficient static transformers. In the United States, the standard power line frequency is 60 Hertz (cycles per second); some countries have standardized on 50 Hertz.

The volt is the quantity of electrical pressure which causes current to flow, and the ampere is the current, or a measure of the number of electrons which pass through a conductor in a given length of time. Ohm's law states that the resistance of a load in ohms is equal to the voltage across it divided by the current passing through it.

The unit of power is the watt, which is equal to the time rate of doing work. One watt is defined as the power dissipated in a resistive load that draws one ampere of current when subjected to one volt of potential (watts = volts times amperes).

The service technician measures voltage with an instrument called a voltmeter; current is measured with an ammeter. Power is measured with a wattmeter.

In AC power systems, the phase relationship of the voltage sine wave and current sine wave is often not identical (exactly in phase). The current drawn by a load may lead or lag the applied voltage by up to 90 electrical degrees. Such a load is called reactive because it contains either capacitive or inductive components. The extent to which the voltage and current are out of phase is depicted by a quantity referred to as the power factor. This is defined as the cosine of the angle of lead or lag, and can be any value from 0 to 1.

When computing power in an AC load, power factor must be taken into consideration. The quantity volt-amperes drawn by any load from a power line is equal to the product of the voltage times the current, but the power dissipated in that load is equal to the volt-amperes multiplied by the power factor.

Power-generating and transmission systems in the United States are designed to provide various configurations of AC power, depending upon the application. For residential use, a nominal 240-volt, single-phase, three-wire system is common, with the center leg of the power source (called neutral) held at ground or earth potential. This system, illustrated in Figure 1.5, has the advantage that a 240-volt AC power source can be supplied to a residence or business, and its maximum potential to ground is half the maximum voltage, or 120 volts. This is an important

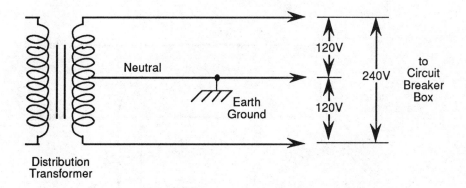

Distribution
Transformer

**Figure 1.5   120/240 Volt Single Phase Power
Supplied to Residential Customers**

safety consideration. High-powered loads, such as central air conditioning systems, are operated from the two outside legs of the system which
provide 240 volts single phase power. Ordinary loads (120 volts) are
powered from either outside line and neutral.

Note that although the nominal voltage rating for single-phase service
in the United States is usually 120 and 240 volts, many air conditioners
are rated for 115 volt or 220 volt operation. This takes into account a
small voltage drop which will occur in the power lines feeding the
appliance.

The single-phase AC systems (Figure 1.5) used in residential service
cannot easily provide large amounts of power which may be required for
industrial applications. For these situations, three-phase power is
predominant.

Three-phase power (Figure 1.6) is a system of transmission which uses
three or four feed wires to provide three discrete voltages, each 120
degrees out of phase with the other two. One advantage, among others,
is that for a given amount of power transfer between two points, the
required amount of copper (or aluminum) wire, by weight, is significantly less. The reason for this is that while the instantaneous current being
delivered by one phase to a load is traveling in one direction, the current
of either or both of the other phases is traveling in the opposite direction.
The result of this is that the currents in each of the wires partially cancel
each other out, resulting in less net current. This allows smaller
diameter wires to be used.

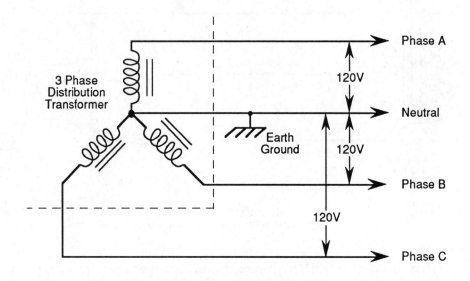

Figure 1.6 120/208 Volt 3 Phase 4 Wire Depiction.
Voltage Between any Two Phase Wires is 208 Volts

An additional advantage is the ease with which three-phase power may be generated. It is produced in alternators which have stator windings which are mechanically offset from each other by 120 mechanical degrees.

Three-phase power is usually supplied to small industrial or commercial customers in a form called 120/208 volt three-phase four-wire wye. The term wye comes about because the graphic illustration of the system resembles the letter Y. The fourth or center wire, called neutral, is at a potential of 120 volts from each of the other three wires and is held at ground or earth potential. This allows ordinary 120 volt AC loads such as light bulbs, small appliances, etc., to be powered by any one leg of the three phases and the neutral wire. When a high-powered device, such as a central air conditioner, is to be operated, it will be powered either by any two legs of the three-phase system (208 volt single phase operation), or by all three legs (208 volt 3 phase operation). Many units are designed for 208/240 volt input, and will operate satisfactorily over that range of voltage.

Higher power three-phase systems are used for very large industrial customers. These generally are rated at higher voltages, such as 440 or more volts line to line, and may not include a neutral wire. Such a three-wire three-phase system is called delta because its graphic notation resembles the Greek letter delta. Distribution transformers, located at the customer's site, provide the lower voltages required for common electrical use.

Three-phase power is especially efficient when used for operating large induction motors, such as air conditioning compressors and blowers. It provides high starting torque without the need for capacitors, and allows the use of physically smaller motor designs. The direction of rotation of three-phase motors is determined by the phase sequence of the power line, and is easily reversed (when desired) by interchanging any two of the three power feed wires.

1.5 Pressure/Temperature Fundamentals

The air conditioning service technician should be thoroughly familiar with the relationship between pressure and temperature as it applies to refrigerants and air conditioning system operation. Pressure is a force per unit area, and rated in pounds per square inch (PSI) in English measure. In the metric system, pressure is expressed in KiloPascals (KPa). One PSI is equal to 6.89 KPa.

Pressure measurements as used in air conditioning service work are read by a gauge that normally reads zero when no connection is made to it, even though atmospheric pressure (14.7 PSI at sea level) is present at the gauge input. The pressure reading indicated by the gauge is correctly referred to as pounds-per-square-inch gauge (PSIG) because it is the actual reading of the gauge, and not the absolute pressure, which would be 14.7 PSI greater. In essence, the common air conditioning service gauge is actually a differential gauge with the reference pressure equal to 14.7 PSI at sea level.

Although the correct nomenclature for the manifold gauge reading is PSIG, it is usually referred to as PSI.

Since the gauge reads zero when no pressure is applied to its input, the concept of negative pressure or vacuum comes into play. If one were to connect the gauge to a vacuum pump that could remove virtually all the air that was contained in the connecting hose, the gauge would read a negative pressure. At sea level, this reading would be -14.7 PSI, because

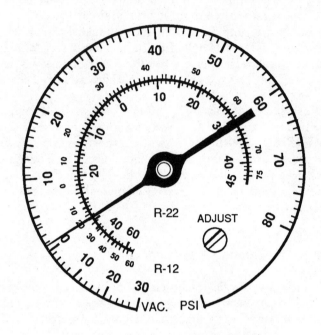

Figure 1.7 Low-pressure Compound Gauge Scale

that is the amount of offset built into the gauge when it is manufactured. In air conditioning work, negative pressure readings are generally not used. Instead, a mercury barometric scale, in which 14.7 PSI is equal to 29.92 inches of mercury, is specified. With the gradual changeover to the metric system, KPa units may be used instead. On the low pressure manifold gauge (Figure 1.7) used by air conditioning service technicians, pressures less than zero are indicated by a scale calibrated in inches of mercury.

For extremely low-pressure measurements, such as might be encountered when dealing with natural gas or propane regulators used in absorption air conditioning systems, the 0-to-29.92-inch mercury scale is too coarse and is replaced with a 0-to-33-feet column of water scale. Pressure measurements using the water column scale are usually less than 1 PSI, which is equivalent to 27.7 inches of water column. Note that a mercury or water column scale may be used for both positive and negative pressure measurements, by using the force of pressure or vacuum to raise the column of liquid.

Air conditioning systems, and some controls such as thermostats and expansion valves, depend upon the relationship of pressure and temperature of liquids and gases in a fixed volume system.

Table 1.1 Pressure/Temperature Chart for R-12 and R-22

Temperature		Freon	Freon
F	C	12	22
–30°	–34.4°	* 5.5	4.8
–25°	–31.7°	* 2.3	7.3
–20°	–28.9°	0.6	10.1
–15°	–26.1°	2.4	13.1
–10°	–23.3°	4.5	16.4
–5°	–20.6°	6.7	20.0
0°	–17.8°	9.2	23.9
10°	–12.2°	14.6	32.7
20°	–6.7°	21.0	43.0
25°	–3.9°	24.6	49.7
30°	–1.1°	28.5	54.8
35°	1.7°	32.6	61.4
40°	4.4°	37.0	68.5
45°	7.2°	41.7	76.0
50°	10.0°	46.7	84.0
60°	15.6°	57.7	101.6
70°	21.1°	70.2	121.4
80°	26.7°	84.2	143.6
90°	32.2°	99.8	169.4
100°	37.8°	117.2	195.9
120°	48.9°	169.1	277.9
130°	54.4°	181.0	296.8
140°	60.0°	206.6	337.2

* = inches of mercury (vacuum)

In a given volume such as a container or sealed air conditioning system, in which both liquid and gaseous refrigerants are present, the internal pressure is determined by a quantity known as saturated vapor pressure. This is the condition of equilibrium in which the liquid refrigerant boils off, at the prevailing temperature, until the pressure in the container is at the point where any additional pressure caused by boiling would cause condensation of the vapor.

In thermostatic controls and expansion valves which use this principle, the sensing bulb or element is partially filled with a mixture of liquid and gas, and is under pressure. As the temperature of the element varies, the pressure within also changes, allowing the device to perform some mechanical function such as operating a switch or changing the position of a needle valve.

The saturated vapor pressure of any liquid is directly proportional to its temperature and is a known and predictable quantity. This parameter is often referred to when diagnosing or servicing a unit. This concept is so important that the scales of the manifold gauge set contain temperature scales for the most common refrigerants. At any given pressure reading, the accompanying temperature reading for the refrigerant under consideration is the boiling point of that refrigerant at that pressure level. Table 1.1 is an abbreviated chart which illustrates the pressure/temperature relationships of refrigerants R-12 and R-22.

1.6 Mechanical Refrigeration Cooling

Refrigeration technology utilizes the laws of thermodynamics and physics to prove a transfer of heat from a lower temperature ambient level to a higher one. When one refers to a refrigerated or cooled area, it really means that heat has been removed, and not "cold has been added." Heat always naturally flows from a warmer environment to a cooler one. In the case of refrigeration, it is desired to reverse this process. This takes energy, or work. Heat is energy, and the law of preservation of energy states that it can neither be created nor destroyed. What we have, therefore, in a refrigeration system, is a "heat pump."

In order to understand the operation of a mechanical refrigeration system, some basic laws of physics should be known. All refrigeration systems which utilize a compressor as the working element provide the desired heat transfer through the action of changing a substance, called a refrigerant, from a gas to a liquid and back to a gas again in a closed loop. The use of a refrigerant in liquid and gas form makes it convenient to transfer heat from one location to another by means of a system of tubes or pipes.

A refrigerant may be any substance which can be conveniently changed from a gas to a liquid, and back again, by means of controlling pressure. Indeed, almost any substance could be theoretically used. But in any refrigeration or air conditioning system, there are certain restraints which limit the number of satisfactory compounds which make a suitable refrigerant. One of the obvious restraints is the relationship between the boiling point and pressure of the refrigerant, which must be at reasonable levels.

In the early days of mechanical refrigeration, sulphur dioxide was used as a refrigerant. This compound could easily be changed from a gas to a liquid at temperatures and pressures which were ideal for a refrigeration system. It was easily obtainable and cheap. However, there was one important disadvantage: It was a harsh, acrid gas which would at the very least cause much human discomfort in the event that it leaked into the atmosphere. Large quantities could prove to be dangerous to human safety.

When the family of Freon refrigerants was invented, it became the panacea for the refrigeration and air conditioning industries. For many applications, Freon 12 and Freon 22 became the refrigerants of choice. These were compounds which were chemically stable, easily manufactured at low cost, 100% odorless and non-toxic, and had the desired temperature/pressure characteristics. At the time that these Freon refrigerants were invented, it seemed that they were the perfect solution for use in mechanical refrigeration systems. Of course, the widespread use of certain Freon compounds, called chlorofluorocarbons (CFCs), is now known to be a threat to the environment, and will be phased out of use by the year 2000.

The pressure/temperature chart (Table 1.1) of common refrigerants indicates the saturated pressure of these compounds as a function of temperature. Conversely, the temperature at which a refrigerant will boil is indicated on the chart, and is determined by the pressure to which it is subjected. By controlling the pressures in a mechanical refrigeration or air conditioning system, the temperature of the boiling liquid refrigerant in the evaporator can be controlled, allowing the liquid to absorb latent heat as it changes from a liquid to a gas at the desired temperature. When the gaseous refrigerant is recycled back to the condenser under high pressure, it gives up its latent heat as it changes to a liquid. This is the basic technique by which mechanical refrigeration

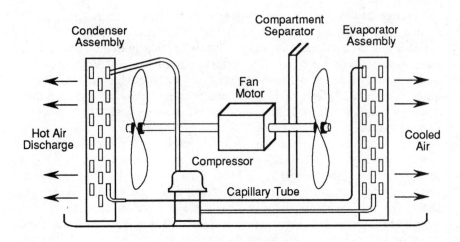

**Figure 1.8 Simplified Illustration of a Compression Cycle
Air Conditioning Unit**

works. A simplified illustration of the components of a typical window
air conditioning unit is shown in Figure 1.8.

A typical compression cycle air conditioning or refrigeration system
can be described as follows:

When the compressor is operating, it takes refrigerant gas at a
relatively low pressure and compresses it to a much higher pressure. In
so doing, the gas temperature rises because its heat energy has been
concentrated into a much smaller volume. This state is called superheated, because the gas is at a higher temperature than the boiling temperature for the existing pressure.

The superheated gas is passed through an assembly of tubing or coils,
called a condenser, which has a relatively large surface area exposed to
the cooler ambient atmosphere. When the gas passes through the air-
cooled condenser it gives up the heat that it had accumulated during
compression and condenses into a liquid.

The liquid refrigerant is then passed through a metering device which
restricts the flow of refrigerant. A simple but effective metering device
found in many refrigeration and air conditioning systems is a length of
small diameter tubing called capillary tubing.

Since the high-pressure liquid refrigerant encounters a restriction when passing through the metering device, the pressure on the exit side of the device is relatively low, and the liquid boils at the temperature indicated in the pressure/temperature chart.

In order for the liquid to change state into gas, it must absorb heat from its surroundings. The liquid refrigerant, passing through a series of coils called the evaporator, causes the temperature of the coils to decrease. The air flow through the evaporator provides the heat necessary to vaporize the refrigerant. When the refrigerant, now in a gaseous state, leaves the evaporator, it returns to the suction side of the compressor where the cycle repeats.

The manufacturer of the refrigeration equipment determines the operating pressures of the system, and thereby the temperatures of the condenser and evaporator. The size of the compressor and other factors determine how much refrigerant will pass through the system in a given amount of time, and thus the BTU rating of the equipment is established.

1.7 Compressors

There are several types of compressors which have been employed for refrigeration and air conditioning service. Possibly one of the first types that was used commercially is the common reciprocating compressor (Figure 1.9), which has a geometry very similar to a one-cylinder gasoline engine. This type of compressor usually was belt-driven by an external electric motor to provide the required mechanical motion. This type of arrangement required that the crankshaft of the compressor be brought out from within the assembly through a seal which prevented leakage of refrigerant. This was one of the reciprocating compressor's main disadvantages.

The disadvantage of shaft seal requirement in a reciprocating compressor was eliminated by the development of the hermetic compressor. This type of compressor has been used for air conditioning and refrigeration systems for many years, and only now is being replaced by newer design compressors which exhibit smaller size, lower cost, and higher efficiency.

The hermetic compressor, sometimes shaped like a dome or pancake, contains the rotor and stator of an induction motor, as well as all the necessary components of the reciprocating compressor. The entire

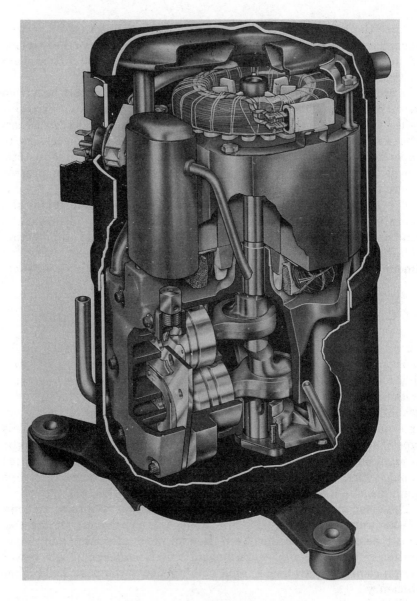

Figure 1.9   Reciprocating Compressor
(Courtesy of Tecumseh Products Co.)

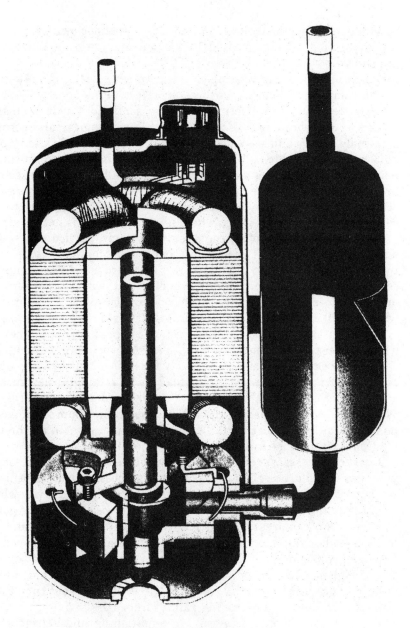

Figure 1.10   Rotary Compressor

assembly is hermetically sealed in a welded steel enclosure with only the connecting tubing exposed to the outside world.   When this assembly is
installed in an air conditioning or refrigeration system, the integrity of the compressor housing ensures a perfect hermetic seal.

Early versions of the hermetic compressor used 4 pole induction motors which operated at about 1800 rpm. These compressors operated extremely quietly and seemed to last forever. Later designs used 2 pole 3600 rpm motors which allowed a size reduction, but were significantly noisier due to the higher speed. The newer, smaller units also seemed to have a higher failure rate than the old designs.   Some hermetic compressors in use today are designed for two-speed operation, which results in lower operating cost when full unit capacity is not required.

Some manufacturers of air conditioning systems (particularly small-sized window units) have used rotary compressors for many years. The trend nowadays for many manufacturers is toward using such compressors in favor of the reciprocating type, since this permits a far smaller and lighter weight design than the hermetic reciprocating compressor which it replaces.   Figure 1.10 illustrates a typical rotary compressor.

The operation of a basic, two-blade rotary compressor is as follows: As a blade passes the suction port, the low-pressure vapor is drawn into the compressor cavity.   When the second blade passes the suction port, it seals the refrigerant between the two blades while starting a new suction cycle behind it. As the blades revolve, the trapped gas is compressed. When the first blade passes the exhaust port, the compressed gas is pushed into the exhaust port. The cycle is thus repeated as the two blades rotate past the intake and exhaust ports of the compressor.

Rotary compressors are not limited to two blades and some commercial units have as many as eight. Some rotary compressor designs utilize a stationary blade system with an eccentric shaft within the cavity which compresses the gas as it travels between the intake and exhaust ports.

A new type of compressor design has recently entered the market. This is the scroll compressor (Figure 1.11) which has been developed by Copeland, a subsidiary of Emerson Electric Company. The scroll design boasts higher efficiency ratios, 12 or more, than can be obtained from either reciprocating or rotary compressors. At the present time these units are available in horsepower ratings from 1 1/2 to 5.

The evaporator is part of every air conditioning and refrigeration system. In air conditioning systems, it usually is composed of a series of copper or aluminum tubing held together with thin sheet metal fins. The

Figure 1.11   Picture of Scroll Compressor
(Courtesy of Copeland)

purpose of the fins is to provide a large surface area which can be exposed to a stream of forced air.

As its name implies, the evaporator is the component of the system into which the liquid refrigerant under high pressure is metered in controlled quantities. As the liquid enters the series of tubes, it changes into a gas (evaporates) since the ambient low pressure in the evaporator cannot sustain refrigerant in a liquid state.

1.8 Evaporator

Evaporators can operate in either of two ways, depending upon the design of the system. A dry-system evaporator receives liquid refrigerant only as fast as needed to maintain the desired temperature. A flooded evaporator always contains some liquid refrigerant, and some of this liquid will find its way out of the evaporator and into the suction line of the system. Flooded evaporators are always followed by a storage tank called an accumulator, which prevents liquid refrigerant from reaching the compressor where it could cause damage.

When the refrigerant changes state from liquid to gas, it absorbs heat from the surrounding metal surfaces. This heat is supplied by the steady stream of warm, humid air which is eventually recirculated back to the area that is to be cooled.

The temperature at which the liquid refrigerant boils is determined by the evaporator or suction pressure of the system. Since this temperature is almost always lower than the dew point of the unconditioned air, water vapor present in the air condenses on the outside of the coils and fins. This is actually a beneficial side effect of air conditioning systems, since in most parts of the country uncontrolled summertime humidity is higher than what would be considered to be a comfortable level.

Note that any dehumidification of the air will absorb BTU energy in the form of latent heat and will depend upon the temperature of the coils, air velocity, and other factors. As a result, the BTUs which are expended in removing moisture from the air will not be available to lower its temperature. For this reason, air conditioning units produced by different manufacturers will remove different quantities of moisture from the air, and can have different inlet/outlet air differential temperatures.

1.9 Condenser

The condenser in an air conditioning or refrigeration system can be very similar to the evaporator in its construction, being composed of a series of coils of tubing, but its function is exactly opposite. Any heat pump which makes use of Freon or other refrigerants depends upon the change of state (liquid and gas) of the refrigerant to exploit the latent heat exchange which is a result of that process. In the case of the condenser, it is the function of that component to provide sufficient surface area to high-pressure, high-temperature gas so that it releases heat to the atmosphere and condenses to a liquid.

When the refrigerant is discharged from the compressor it is in a superheated state. Superheat can be defined as the temperature of a gas which is above its boiling point as a liquid at the existing pressure, as illustrated in a pressure/temperature chart. Since the high pressure gas refrigerant enters the inlet side of the condenser at a relatively high temperature, the heat energy present in the gas is dissipated to the atmosphere. Heat always travels from a higher temperature medium to a lower one. Air conditioning units always employ a forced air or water system which removes the desired quantity of heat from the refrigerant as it travels through the coils of the condenser. Refrigeration systems may also use forced air, but many household appliances can operate satisfactorily with only the natural convection of air to provide the necessary cooling.

When the refrigerant present in the condenser gives up its heat to the atmosphere, the temperature of the fluid falls below the point where it can remain a gas at its existing pressure. As a result, the refrigerant is condensed and reaches the condenser outlet in a totally liquid state. The temperature of this liquid is not very significant, but it will be somewhat close to ambient temperature.

1.10 Capillary Tube

As indicated in the simplified diagram of a mechanical refrigeration system, there must be a point in the system which separates the high-pressure side (condenser) from the low-pressure side (evaporator). This function may be performed by one of several different components, but the end result is similar.

Separating one pressure level from another can be performed by placing some kind of restriction in the line feeding the two sections. One might think of this in terms of an ordinary garden hose where a sharp bend in the hose restricts the flow of water. The pressure before the bend will be higher than the pressure past the restriction. The same analogy holds true for the restricting device used in an air conditioning or refrigeration system.

One of the simplest types of devices to perform this function is called a capillary tube, which is a length of fine bore tubing that controls the flow of liquid refrigerant as it passes from condenser to evaporator. Most domestic refrigeration systems and many window-type air conditioning units use the capillary tube as the refrigerant metering device. Figure 1.12 illustrates a typical capillary tube which finds application in a compression cycle refrigeration system.

The capillary tube has several advantages which makes it an ideal refrigeration control. It has no moving parts to wear out or stick. It is extremely cheap to manufacture. Lastly, it permits the high-side system pressure and low-side system pressure to equalize when the unit is shut down between operating cycles. This permits the use of more economical lower starting torque compressors.

The capillary tube works on a very simple principle—friction. It is easy to understand that a very large diameter tube will allow a greater flow of liquid (given constant pressure) than one of smaller diameter. Moreover, the amount of total friction can be easily controlled, with a given size tubing, by specifying its length. Capillary tubes are designed with a very small inside diameter, possibly as small as .031 inch, or one millimeter. The diameter and length of the capillary tube is determined by the manufacturer of the unit in accordance with the BTU rating of the equipment.

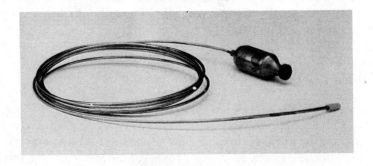

Figure 1.12 Typical Replacement Capillary Tube
Showing Filter Screen Assembly

Since the inside diameter of the capillary tube is very small, it is mandatory that no dirt enter the tube, which will clog it and render the system inoperative. To prevent such a situation, capillary tubes are often preceded by a small filter assembly.

Well-designed refrigeration and air conditioning systems that use capillary tubes will have the tube mechanically bonded to the outlet pipe of the evaporator, which often is at a lower than ambient temperature. Additionally, many window-type air conditioning units route the capillary tubes through the trapped cold water which rests in the bottom of the unit. The thermal transfer of heat from the warm capillary tube to the colder evaporator tube or water helps ensure that any trapped refrigerant gas in the capillary tube is condensed to liquid before it reaches the evaporator. This small design feature will wring a few more BTUs from the hard working air conditioning or refrigeration system, and enhance its EER.

1.11 Expansion Valves

A more complicated but higher performance restricting device used in air conditioning systems is the expansion valve. This family of refrigerant controls includes many forms, each with its particular advantages. They are automatic in operation, controlling the flow of liquid refrigerant in accordance with the cooling needs of the system. Some expansion valve designs permit adjustment in the field to tailor valve performance to the particular equipment and cooling load.

The automatic thermostatic expansion valve (Figure 1.13) regulates the flow of liquid refrigerant in accordance with the temperature of the evaporator coils. When the air conditioning system is first turned on the evaporator coil is at a relatively high temperature. This is sensed by the expansion valve bulb which is mechanically attached to the evaporator outlet tube. The pressure developed by the liquid/gas mixture within the sensing bulb is transmitted to the diaphragm through a capillary tube, causing the valve to open against the closing action of the internal spring. This permits maximum flow of liquid refrigerant into the evaporator, providing maximum cooling. As the temperature of the evaporator outlet tube decreases, the lower pressure applied to the diaphragm from the sensing bulb assembly is partially overcome by the spring, causing less refrigerant to flow. The expansion valve thus reaches equilibrium as the flow of liquid refrigerant into the evaporator is just sufficient to meet its cooling needs.

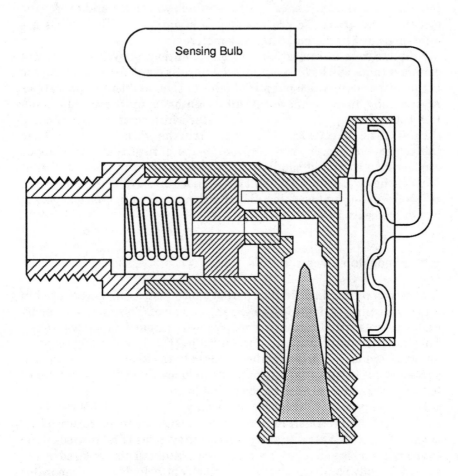

Figure 1.13 Thermostatic Expansion Valve

Evaporator temperature sensing is usually accomplished by the use of a small metal bulb which is filled with a liquid and/or gas which may or may not be the same as the refrigerant used in the system. The bulb assembly is mechanically and thermally connected to that part of the evaporator for which the temperature is to be sensed (usually near the evaporator outlet). In this way, the valve automatically controls the flow of liquid refrigerant.

The thermostatic expansion valve does not know anything about the condition of the air conditioning system except the temperature of the evaporator coils. As these coils get colder, the valve automatically restricts the flow of refrigerant further, resulting in even lower evaporator pressure. In a malfunctioning system, where the flow of warm evaporator inlet air is restricted (due to a clogged air filter, for example), the resulting closure of the valve can result in abnormally low evaporator pressure and freezing of the coils.

1.12 Fixed Expansion Orifice

A newer type of refrigerant control device has found its way into many automotive air conditioning systems. Similar devices may also be found in newer central air conditioning systems. This is the fixed orifice (Figure 1.14), which operates in a similar manner as the capillary tube in a residential A/C unit. A fixed orifice has no moving parts, which makes it a very low-cost, reliable control. It is composed of a filter screen and tube with a precise inside diameter so that a known quantity of refrigerant will pass through at a given pressure differential between input and output. On AC-operated air conditioning systems, constant compressor speed and limited ambient temperature conditions allow a properly designed fixed expansion orifice to be used without the possibility of an excessively cold evaporator and frost build-up.

The fixed orifice control does not have any way of knowing the instantaneous operating condition of the automotive air conditioning system, which can vary over a wide range due to constantly changing compressor rpm and heat load on the evaporator. As a result, there will be a condition where the suction pressure of the compressor (and evaporator) will cause the refrigerant boiling temperature to fall below the freezing point of water. This will cause ice to form on the evaporator coils, inhibiting virtually all cooling. This problem is solved by using a pressure or temperature switch that monitors evaporator operating conditions and cycles off the compressor when the evaporator gets too cold.

One of the big advantages of using the fixed orifice control in an automotive air conditioning system is the fact that the compressor is required to operate only part of the time in accordance with the heat load. This saves precious fuel over the older designs using traditional expansion valves for refrigerant control.

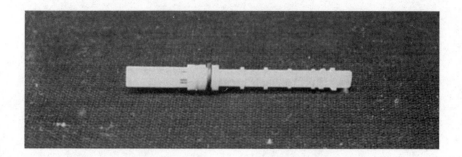

Figure 1.14   Typical Fixed Expansion Orifice

When used as part of a central air conditioning system, the fixed expansion orifice provides a low-cost, efficient method of regulating liquid refrigerant flow.

1.13 Evaporator Pressure Control

Evaporator pressure controls are used in some automotive air conditioning systems to maintain a minimum evaporator pressure, usually not much less than 30 PSI for systems using R-12 as a refrigerant, to prevent evaporator coil freeze-up under any operating condition. By preventing the evaporator pressure from going much below 30 PSI, the temperature of the coils is held above the freezing level. The evaporator pressure control valve is placed between the suction inlet of the compressor and evaporator outlet tube.

One type of evaporator control is called a pressure operated absolute (POA) valve, illustrated in Figure 1.15, which contains a bellows which is totally evacuated. The pressure within the bellows, essentially 0 PSI absolute, is the reference against which the POA valve operates. This valve is immune to changes in atmospheric pressure.

Other types of evaporator pressure control valves operate with a gas-filled bellows. This type, and the POA valve, monitor evaporator pressure and automatically compare this pressure to the force of a spring which tends to hold the valve closed. When evaporator pressure exceeds the predetermined level, the valve opens to allow greater refrigerant flow.

These types of valves were widely used in many General Motors and Chrysler automotive air conditioning systems to ensure that the evaporator

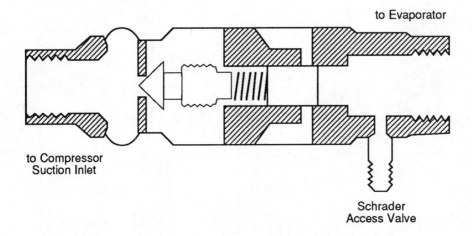

Figure 1.15   Simplified View of POA Valve

pressure did not go below the level which would cause condensed water to freeze on the evaporator coils. This feature is particularly important in automotive air conditioning systems which are often operated in all four seasons of the year.

1.14 Receiver/Drier

A receiver/drier (Figure 1.16) is a storage container for liquid refrigerant that is placed in the liquid line between the condenser and expansion valve. It also serves as a location to store a desiccant (drying medium) which absorbs any moisture which may be in the system.

The use of a receiver in an air conditioning system is required for most systems which employ an expansion valve for refrigerant control, since it provides a location for liquid refrigerant to accumulate should the expansion valve slow down the flow. Those systems which use capillary tubes do not require a receiver, since the capillary tube can never shut down.

Liquid refrigerant enters the receiver/drier at the top of the container, and leaves it through a tube which reaches down to the bottom. This ensures that only subcooled liquid refrigerant and no gas is delivered to the expansion valve.

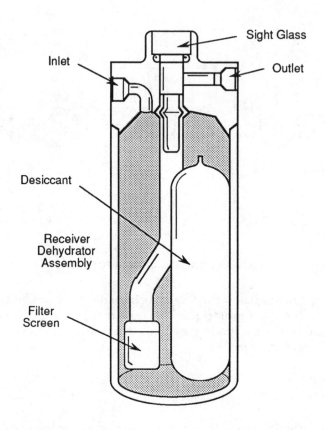

Figure 1.16 Receiver/Drier

1.15 Accumulator

Some air conditioning systems require the use of an accumulator (Figure 1.17), which is a protective component that is used as a storage container for liquid refrigerant. It is usually located in the vapor line between the evaporator and compressor suction inlet port. This is particularly important for automotive air conditioning systems which employ an expansion orifice for refrigerant control, and operate with a "flooded"

evaporator. Many residential and commercial units also employ accumulators to prevent liquid refrigerant from reaching the suction inlet of
the compressor, which can cause slugging.

Most AC-powered air conditioning systems operate with a dry evaporator (all liquid refrigerant is vaporized), and normally the use of an
accumulator is not required. However, the manufacturer of the system
sometimes includes an accumulator as a conservative measure to ensure
compressor protection under any condition of operation.

As with the receiver, the accumulator can also be used as a location to
store the desiccant.

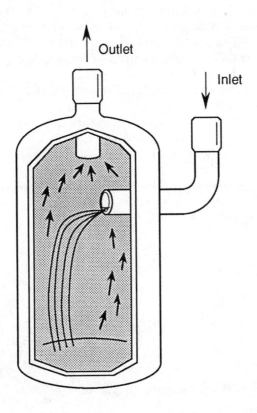

Figure 1.17 Accumulator

1.16 Thermostats

Most air conditioning systems employ a thermostat to allow the system to shut down automatically when the desired room temperature has been reached. Small window-type units usually are equipped with a thermostat which contains a heavy-duty switch that controls power to the compressor. Newer models contain "energy saving" thermostats which have an additional set of contacts that allow the fan motor to be automatically cycled on and off with the compressor. These types of thermostats operate at the line voltage, 120, 208, or 240 volts.

Such thermostats may utilize either a bimetallic strip of metal which bends as temperature changes, or may use a sensing bulb or capillary tube that is filled with a refrigerant. As the ambient temperature changes, the pressure within the closed element exerts a mechanical force which operates a switch. Figure 1.18 illustrates a typical capillary-tube-controlled thermostat.

Central and commercial air conditioning systems usually employ a cooling or combination heating/cooling thermostat which is operated in a 24-volt AC circuit. This type of system controls a relay or contactor which in turn allows power to the compressor to be controlled.

Larger systems may be designed for two-stage operation, where the thermostat controls two compressors. This type of operation is accomplished by having two set points on the thermostat, separated by about two degrees. This allows staggering of compressor operation in accordance with the cooling needs of the controlled area.

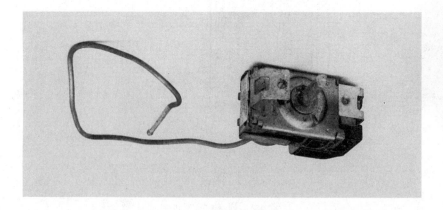

Figure 1.18 Thermostat with Sensing Capillary Tube

1.17 Pressure Controls

Central and commercial air conditioning systems generally employ one or more pressure controls to monitor system operation and shut it down in the event of a malfunction, which causes an abnormal pressure condition. Excessive compressor head pressure and very low suction pressure are two examples of conditions which could cause compressor damage.

Failure of the condenser fan motor or blocked condenser fins are two possible causes of excessive compressor head pressure. Partial or total loss of the refrigerant charge will be a cause of low suction pressure. The high-pressure cutout generally is designed to be manually reset so that the system does not automatically resume operation when the pressure returns to normal. Low-pressure cutouts may be automatic reset types.

1.18 Solid State Control Modules

Many air conditioning units manufactured today are equipped with solid state controls and relays which have taken over the function of the mechanical parts of the past, such as switches and thermostats. It is not uncommon to find a complex printed circuit board as part of the control circuitry of even the smallest window-size air conditioner.

The electronic circuitry, which usually includes integrated circuits or a microprocessor, provides all the user control functions (blower speed, thermostat, etc.) as well as automatic time delay circuits to prevent short-cycling. The heavy current demanded by the compressor and blower motor is switched on and off by means of relays. A typical solid state control module is shown in Figure 1.19.

Although failure of even the most inexpensive part on such control modules can render the air conditioner inoperative, most service technicians do not have the expertise to diagnose the problem down to the component level. In this case, replacement of the entire module may be required.

Figure 1.19 Typical Solid State Control Module
Used in a Window Air Conditioner

Air Conditioning Heat Loads

2.1 General Information

Any air conditioning system must be designed so that it is capable of removing the quantity of heat which enters the controlled area under the worst case conditions, such as occur with high-temperature, high-humidity days. Additionally, all heat-generating factors, such as human occupancy, lighting, electrical equipment, and any other energy-consuming devices must be accounted for when calculating the maximum heat load of a given area. Consideration must also be given to latent heat-generating factors, such as dishwashers and laundry appliances, which emit water vapor that must be removed by the air conditioning system.

The highest temperature that will be experienced in any given area of the United States can be obtained from reference data books published by the American Society of Heating, Refrigeration, and Air Conditioning Engineers (ASHRAE). Such charts, illustrated in Table 2.1, provide historical information on the prevailing temperatures that can be expected during the cooling season.

Since the extremes of temperatures illustrated in the chart are of short duration, an air conditioning system which is designed for the maximum temperature listed may be too large in capacity. Excessive cooling capacity should be avoided, since it will result in a system in which the cooling cycle will be too short, and during the off time the humidity buildup in the controlled area will become uncomfortable. A better design is to use a smaller size unit in which the compressor operates most of the time, maintaining the dehumidification process. Oversize units also have other disadvantages such as higher initial and operating costs.

A common method to calculate the required size of air conditioning equipment is to design for a given differential between outside and inside temperatures. For example, if the normal daytime summer ambient temperature is usually about 90 degrees F (32 degrees C), a system designed for 15 degrees F differential will be able to maintain 75 degrees F (24 degrees C) inside when running continuously.

With such a system, the conditioned air will always be maintained at the design level of 75 degrees when the outside temperature is 90 degrees or less. Should an out-of-the-ordinary heat wave occur where the outside temperature hovers around 100 degrees, for example, the inside will rise to 85 degrees. This might not be as terrible as it sounds, since the dehumidification of the inside air will help maintain comfort during the extraordinary cooling load on the system.

Table 2.1   Chart Illustrating Average and Peak Summer Temperatures throughout the United States

City	Average July Temperature	Peak Temperature
Atlanta GA	78 F 26 C	103 F 39 C
Atlantic City NJ	72 F 22 C	104 F 40 C
Boston MA	72 F 22 C	104 F 40 C
Burlington VT	70 F 21 C	100 F 38 C
Charlotte NC	78 F 26 C	103 F 39 C
Chicago IL	74 F 23 C	103 F 39 C
Cincinnati OH	75 F 24 C	105 F 41 C
Denver CO	72 F 22 C	105 F 41 C
Detroit MI	72 F 22 C	104 F 40 C
Indianapolis IN	76 F 24 C	106 F 41 C
Jacksonville FL	82 F 28 C	104 F 40 C
Milwaukee WI	70 F 21 C	102 F 39 C
Nashville TN	79 F 26 C	106 F 41 C
New Orleans LA	82 F 28 C	102 F 39 C
New York NY	74 F 23 C	102 F 39 C
Omaha NE	77 F 25 C	111 F 44 C
Portland OR	67 F 19 C	104 F 40 C
Saint Louis MO	79 F 26 C	108 F 42 C
Salt Lake City UT	76 F 24 C	105 F 41 C
San Francisco CA	58 F 14 C	101 F 38 C
Seattle WA	63 F 17 C	98 F 37 C
Washington DC	77 F 25 C	106 F 41 C
Wichita KS	79 F 26 C	107 F 42 C

2.2 Temperature Scales

The most familiar temperature scale in the United States is the Fahrenheit scale, named after a German physicist born in 1686. This scale was devised as 0 being the temperature of an equal mixture of common salt and snow, the freezing point of water at 32 degrees, and the boiling point at 212 degrees at standard pressure. The abbreviation for Fahrenheit is F.

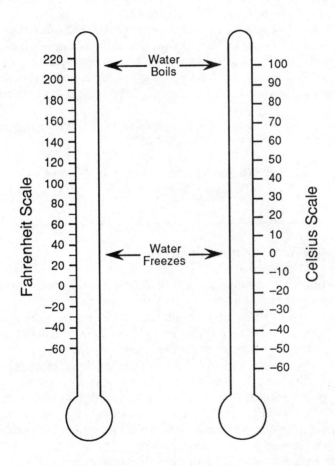

Figure 2.1   Comparison of F and C Scales

In the scientific community a different scale is used, named after the Swedish astronomer Anders Celsius. This scale, sometimes referred to as Centigrade, has its zero temperature as the freezing point of water and 100 as the boiling point. The Celsius or Centigrade scale is abbreviated C.

Another scientific scale is the Kelvin scale which has the same gradient as the Celsius scale, but with its zero defined as absolute zero, −273 degrees C or −459.4 degrees F. Absolute zero is the temperature at which a substance would be totally without heat. The freezing point of water in the Kelvin scale is 273 degrees K.

In air conditioning and refrigeration work, the Fahrenheit or Celsius scale (Figure 2.1) is used. Since the total number of degrees between the boiling and freezing points of water is 180 for Fahrenheit and 100 for Celsius, and is a linear relationship, easy conversion of any temperature from one scale to the other is possible.

To convert Fahrenheit to Celsius, subtract 32 and multiply by 5/9:
Celsius = (Fahrenheit − 32) * 5/9

To convert Celsius to Fahrenheit, multiply by 9/5 and add 32:
Fahrenheit = (Celsius * 9/5) + 32

2.3 Relative Humidity Measurement

Relative humidity is a measure of the quantity of moisture contained in the air, expressed as a percentage of the maximum quantity of water vapor that can exist in that air at that temperature. The maximum possible level of relative humidity is 100%. Any additional moisture above that point will result in condensation or precipitation.

Humidity is closely related to human comfort as is temperature, and any air conditioning system when properly designed will help provide relief from excessive humidity as it does from high temperatures. The temperature/humidity or comfort index, used by many weather forecasters, takes into account the combination of both temperature and relative humidity as it relates to human comfort.

Higher ambient temperatures can accommodate a greater quantity of water in the air than lower temperatures. Because of constant air infiltration in all structures, the warm, moist air which seeps into a cool, controlled area during the cooling season tends to raise the level of relative humidity. Without humidity control, this could prove to be uncomfortable for many people. Additionally, many devices such as photocopy machines and sensitive electronic equipment can suffer in performance if the relative humidity level is not controlled within certain limits.

There are several methods by which relative humidity can be measured, using a family of instruments called hygrometers. One of the lowest-cost instruments relies on a strand of hair, which changes in length in accordance with the amount of moisture that it absorbs. This type of instrument is usually not accurate, especially at extremes of relative humidity levels.

Some electrical methods of measuring relative humidity use a sensor that changes its characteristic as the level of relative humidity varies. Many of these instruments are very accurate, but they invariably come with a high purchase price.

A psychrometer is a low-cost, accurate instrument which is used to measure relative humidity. This device consists of a pair of identical thermometers, one of which has a continuously moistened cloth covering its bulb. One form of this instrument is called a sling psychrometer because in use it is swung through the air to cause a rapid evaporation of the water surrounding the moistened bulb.

Since the water surrounding the wet bulb thermometer gives up latent heat as it evaporates, the indicated temperature is lower than that of the dry bulb thermometer. The rate of water evaporation is directly related to the relative humidity level. By comparing the readings of the two thermometers, a very accurate measurement of relative humidity can be made.

Relative humidity is a measure of the percentage of moisture present in a given volume of air compared to the maximum amount that can be contained in the same volume of air at the same temperature. For this reason, the dry bulb thermometer is used to measure the ambient temperature when the relative humidity measurement is being taken.

A psychrometric chart (Figure 2.2) is supplied with the wet and dry bulb psychrometer. This allows the user to determine the relative humidity measurement by comparing the dry bulb temperature with the wet bulb temperature against the chart. The beauty of this type of instrument is in its simplicity and accuracy, and it is a valuable addition to the service technician's assortment of instruments.

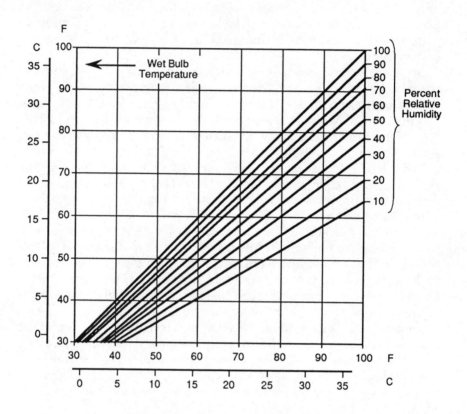

Figure 2.2 Psychrometric Graph

2.4 English and Metric Units

It is important to be familiar with the relationship between English and metric units. The United States is gradually moving toward the metric system for all measurements. It is not uncommon for many air conditioning units and components to be imported, and invariably these will use the metric system exclusively for all mechanical specifications.

The following conversion factors provide a correlation between the more common English and metric units, allowing conversion from one to the other using the multiplying factors indicated. An extensive list of conversion factors can be found in Appendix C.

1. Linear measurement

   To convert feet to meters multiply feet by 0.305
   To convert meters to feet multiply meters by 3.28

2. Area measurement

   To convert sq. ft. to sq. meters multiply sq. ft. by 0.093
   To convert sq. meters to sq. ft. multiply sq. meters by 10.75

3. Volume measurement

   To convert liters to cubic inches multiply liters by 61
   To convert cubic inches to liters multiply cubic inches by 0.016

4. Weight measurement

   To convert kilograms to pounds multiply kilograms by 2.2
   To convert pounds to kilograms multiply pounds by 0.455

5. Pressure measurement

   To convert KPa to PSI multiply KPa by 0.145
   To convert PSI to KPa multiply PSI by 6.89

6. Specific heat measurement

   To convert BTU/lb.F to KJ/Kg.K multiply BTU/lb.F by 4.187
   To convert KJ/Kg.K to BTU/lb.F multiply KJ/Kg.K by 0.239

7. Latent heat measurement

   To convert KJ/Kg to BTU/lb multiply KJ/Kg by 0.43
   To convert BTU/lb to KJ/Kg multiply BTU/lb by 2.3

2.5 Heat Loads

The air conditioning system must be capable of removing the total
quantity of heat that enters the controlled area, or any that is generated

internally. Each of these heat loads must be considered individually, and the total then becomes the required BTU capacity of the air conditioning system.

Any significant heat load, sensible or latent, that will be encountered in a given situation must be considered when calculating the proper size air conditioning unit. Some of these are the more obvious, but others may not be so. The air conditioning system must be able to handle the total heat load, regardless of the source.

Additionally, most air conditioning systems are designed to remove heat from a controlled area only as fast as it enters or is internally generated. No allowance is made for the unit to reduce inside air temperature and humidity to the design level should the system be turned on, during a sustained heat wave, after the controlled area has reached a temperature close to that of the outside ambient. For this reason, air conditioning thermostats should always be set to automatically turn the equipment on when the inside temperature exceeds the desired level.

2.6 Sensible Heat Formula

A simple relationship exists between air flow, temperature rise, and BTU/hr. for heat exchange between a heating or cooling coil and air flow. This is called the sensible heat formula, because all the heat energy represented by the term BTU/hr. is used to raise or lower the temperature of the air, and none of it is consumed as latent heat.

$$\text{SENSIBLE HEAT IN BTU/HR.} = (1.08) * (CFM) * (T1-T2)$$

where cfm equals the total air flow in cubic feet per minute, and T1–T2 is the temperature difference (in degrees F) between the inlet air and outlet air passing through the heat exchanger.

A widely accepted minimum figure for air flow in an air conditioning system is 400 cfm per ton (12,000 BTU/hr.) of cooling. If these figures are applied to the sensible heat formula, the change in temperature across an evaporator coil operating at 400 cfm with 1 ton of cooling is 27.8 degrees F (15.4 degrees C). This temperature differential indicates that at a room ambient of 75 degrees F (24 degrees C), the evaporator air temperature would be 47.2 degrees F (26.2 degrees C).

The same formula may be applied to heat pump systems when operating in heating mode. In this case, the required air flow can be less than 400 cfm per ton, since the temperature differential in heating mode is often significantly greater for heating as compared to cooling.

2.7 Building Insulation Qualities

Probably the greatest heat load encountered by an air conditioning system is conductive heat (sometimes called heat leakage) which is transmitted from outside to inside through walls, windows, ceiling, and floor. The construction of the building will have a large effect on the amount of heat that will pass into the controlled area, and is mostly a function of the insulating qualities of the construction.

A popular method to compute heat transfer through various types of building construction materials is using the "R" or thermal resistance factor, which can be specified by the manufacturer of the material or determined by consulting air conditioning and heating reference manuals. Table 2.2 is a representative chart which illustrates the approximate R value for several types of building materials.

The chart specifies the R value for each component in the composite building wall, ceiling, or floor, and also the insulating effect of air. To obtain the total value of thermal resistance, the R value of each component is added. For example, a brick veneer wall with 1/2-inch wood siding and 3/8 inch sheetrock will have an R value of 0.4 + 0.9 + 0.3, or 1.6. Added to this figure would be any additional insulating qualities provided by insulation, still air, or any other significant thermal resistance factor.

Table 2.2   Typical R Values for Common Building Construction

Type of Construction	R Value
3/8" (1 cm) gypsum wallboard	0.3
1/2" (1.3 cm) gypsum wallboard	0.4
4" glass wool batt	12.0
8" (20 cm) concrete block	1.6
1/2" (1.3 cm) wood siding	0.9
Face brick	0.4
Asphalt shingles	1.0
Still air surface	0.7
4" (10 cm) still air space	1.0

The R value for any given type of construction is the reciprocal of a quantity called U. Thus

$$U = 1/R$$

The expression U, in terms of BTU/sq. ft./deg. F/hour, is the overall heat transfer constant and is used to calculate the total heat leakage, in BTU per hour, of that part of the construction under consideration. To make the calculation, two additional factors must be considered.

The square foot area of the wall, ceiling, or floor becomes part of the calculation, since the quantity U is in the dimension of 1 square foot. Similarly, the differential temperature, in degrees F, between each side of the construction must be included in the calculation.

Thus, for the brick wall described above, if the area was 160 square feet and the design differential temperature was 15 degrees F, the heat loss for that wall would be calculated by multiplying the square foot area by the temperature differential, and dividing by R:

$$\text{Heat leakage} = (160 * 15)/1.6 = 1500 \text{ BTU/hour}$$

2.8 Window Exposure Heat Load

The calculation for walls and ceilings described above does not take into consideration any windows or doors which may be present in the structure. Glass areas typically will have heat leakage possibly four times as great as wall or ceiling construction. A separate chart of R values may be consulted for these, as illustrated in Table 2.3. When calculating wall heat leakage, the area of the windows and doors should be subtracted from the wall area, since the BTU/hour leakage for these components should be calculated independently.

The effect of sun exposure on windows presents an additional heat load which is caused by radiation of heat energy into the controlled area. This factor, calculated in accordance with the total area of any given window,

Table 2.3   Typical R Values for Window Construction

Type of Glazing	R Value
Single glass	0.9
Double glass spaced 1/2" (1.3 cm)	1.54
Triple glass spaced 1/2" (1.3 cm)	2.79
Storm window	1.79

Table 2.4   Radiation Heat Absorption Through
Window Exposure to Sunlight

Direction of Exposure	Total Heat Absorption BTU/Hour/Sq. Ft.
East	55
South	75
Southwest	110
West	100

must be added to the total heat load of the air conditioning system. The total BTU energy is independently calculated for each window, and is a function of the area and direction of sunlight. This will be affected by the geographical location of the building, but a typical chart (Table 2.4) may be consulted to provide a reasonable estimate of the load. Sun heat load on any glass area can be significant on windows facing west or southwest, and any heat reducing methods such as awnings or special glazing containing heat reflective capability will help reduce the total heat load to the system.

Skylights may also have some effect on the total heat load through radiation, depending upon the direction of exposure. These may add as much as 100 BTU or more per hour per square foot, and should be considered when calculating total air conditioning heat loads.

2.9 Internal Heat Generation

All heat generating components within the controlled area must be considered when calculating the total heat load to be placed on the air conditioning system. These include electrical equipment such as lighting, appliances and motors, consumption of natural or propane gas such as during cooking, water heating, or industrial processes, and presence of humans (or animals) which can add a significant amount of BTU energy to the total heat load.

Virtually 100% of the energy drawn by electrical equipment and lighting is converted into heat. This is illustrated in Table 2.5. Energy drawn by electrical devices is specified in watts, and can easily be converted to BTU by multiplying by the factor 3.4. For example, if the total amount of wattage used for lighting in an air conditioned room is 500, the resulting heat generated by the lamps is equivalent to 1700

BTU. Thus, the air conditioning system must remove 1700 BTU per hour to negate the heating effect of the lighting.

Table 2.5 Typical Heat Load Generation

Heat-Generating Component	BTU/Hour Heat Load
Human occupancy (inactive, per person)	400
Human occupancy (extremely active)	2000
1/4 HP motor	1200
Electric stove (per KW)	3413
Natural gas stove (per cu. ft. of gas)	1050
Lighting (per KW)	3413
Appliances (per KW)	3413

When calculating total heat load, any significant increase of humidity (water vapor) within the controlled area must be considered. Water added to the atmosphere will be a result of operation of laundry appliances, automatic dishwashers, and much human activity such as showers, baths, etc.

The total amount of water vapor generated will add an additional load to the air conditioning system in the form of latent heat of condensation. Each pound of water vapor that is added to the inside atmosphere will require 970 BTU of heat energy to condense it to liquid.

2.10 Air Infiltration

An additional heat load that should be considered is air infiltration. All buildings have some amount of air leakage, and this is actually an important and desirable parameter in building construction, since it provides a constant supply of fresh air. Infiltration is the natural transfer of outside air to inside, or vice versa, due to pressure differentials. It is beneficial to some extent because it provides the necessary fresh air that human occupancy requires.

The total amount of air exchange between outside and inside may be 1 or 2 changes per hour. Some buildings, which are tightly constructed with weather-stripped doors and windows and vapor barriers, will have considerably less infiltration. During the cooling season, air infiltration will not only bring in air at outside ambient temperature, but will also increase inside humidity against which the air conditioning system must work.

A rule of thumb which may be used to calculate the required BTU energy to negate the effect of air infiltration is to allow about 5 BTU per hour for each cubic foot of controlled space, assuming a 15 degree F (8.3 degrees C) temperature differential. This factor may be increased if the air conditioned area has greater than average air leakage, such as might be caused by large doorways which are frequently opened. Similarly, the multiplier factor may be reduced for those structures which are designed with extremely tight construction.

2.11 BTU Cooling Load Estimate

The information presented above is often applied in a "rule of thumb" type format in which an air conditioning contractor can compute the correct size cooling equipment for a given situation, whether it be a single room or an entire structure. Basically, this consists of analyzing each component of the total heat load and determining its cooling requirement in BTU per hour. The following outline of such a procedure will provide a relatively easy method to arrive at a reasonable estimate of the correct air conditioner size. Heat pump installations require a more complex analysis of heating and cooling loads to ensure optimum selection of equipment size.

The following method of calculating air conditioning BTU requirements assumes that the temperature differential between outside and inside is held to 15 degrees F (8.3 degrees C). Since the required cooling capacity of an air conditioning system is directly proportional to temperature differential, the multiplier constant specified in the calculations below may be modified by the ratio of the desired differential in degrees F divided by 15.

   1. Wall heat load

South and west, exposed to sun	_____ sq. ft. * 10 =	_____
North and east exposure	_____ sq. ft. * 6 =	_____
Thin walls, any exposure	_____ sq. ft. * 12 =	_____
Interior walls	_____ sq. ft. * 5 =	_____
Interior glass	_____ sq. ft. * 12 =	_____

2. Window heat load

Western exposure   _____ sq. ft. * 75 =_____
Southern exposure   _____ sq. ft. * 50 =_____
Eastern and northern exposure   _____ sq. ft. * 20 =_____
Sun exposed with awnings   _____ sq. ft. * 45 =_____

3. Ceiling heat load

Uninsulated roof above   _____ sq. ft. * 25 =_____
Insulated roof above   _____ sq. ft. * 10 =_____
Second story above   _____ sq. ft. * 4 =_____

4. Floor heat load   _____ sq. ft. * 4 =_____

5. Air infiltration load   _____ cu. ft. * 5 =_____

6. Human occupancy   _____ people * 500 =_____

7. Electrical load   _____ watts * 3.4 =_____

Any additional heat load, such as internal humidity gain or extraordinary heat from any source, should be considered if it increases the total heat load, as computed by the factors specified above, by 10% or more. When all of the heat load factors are taken into consideration, the total becomes the calculated size of the required air conditioning system, in BTU per hour.

If it is desired to calculate the required air conditioner size in tons of cooling, simply divide the total BTU requirement by 12,000. This will be the size of the unit. For example, if the required total BTU capacity calculated above was 60,000, a 5 ton unit would be specified.

Since the calculated heat load will almost always fall between available unit sizes, the air conditioning technician will have some leeway in selecting the unit to be installed. This should be discussed with the customer, who can provide some input, in accordance with economic and/or personal factors as to the selection of the size of the equipment.

In many cases, it may be advisable to choose a unit which is slightly smaller than the calculated size. When this is done, the unit will work at 100% duty cycle during periods of high outside ambient temperatures. The continuous dehumidification of the air will provide a large degree of human comfort, and the smaller size unit will be more efficient and cost effective.

If a larger-than-calculated unit is installed, it has the advantage that it will be able to bring down inside temperatures faster. However, installing a unit larger than required will result in higher installation and operating costs and relatively long off-cycle times. The resulting humidity buildup during the off-cycles can prove to be uncomfortable.

2.12 Energy Efficiency Ratio

Before the oil crisis of the 1970s, not much attention was given to the world supply of energy resources. Such energy was cheap and it seemed that the supply was inexhaustible. As everyone now knows, this is no longer true.

Air conditioning manufacturers did not attempt to provide the most efficient units that the level of technology would permit. The important factor was price. As a result, air conditioning systems built at that time were relatively inefficient.

By law, any air conditioning system sold for household use must carry an energy efficiency rating, called EER. Furthermore, restrictions on the lowest level of efficiency of units that may be legally sold are in effect in certain states. All units must carry a tag indicating its EER, or provide the BTU and wattage rating so that the EER may be calculated. Some manufacturers use the acronym SEER (Seasonal Energy Efficiency Ratio) when specifying the efficiency of their units. For air conditioning units, these two efficiency ratings are identical.

The EER of an air conditioning system is computed as its total BTU output per hour divided by the electrical power draw in watts. For example, a small household window air conditioner with a cooling capacity of 5000 BTU might draw 600 watts from the line. The EER for this unit is then

$$EER = 5000/600 = 8.33$$

Note that the quantity watts, not volt-amperes, is used as the divisor in the above expression. If one wanted to compute EER using the factors of voltage and current, the power factor of the unit must be known, since watts = voltage * current * power factor. The power consumption, in watts, of a window-type air conditioning unit is usually specified by the manufacturer and can be found on the nameplate.

The EER of an air conditioning system depends upon many factors, including the compressor, fan motor, condenser, and evaporator. It is obvious that if a manufacturer wishes to make a unit with the lowest possible manufacturing cost, all of these components will be designed with price as the determining factor, not performance.

If one was to compare a low-efficiency window air conditioner with an EER of 6 (millions of these were produced) with a modern unit that has an EER of 10 or more, the most obvious difference that will be noted is the physical size of the unit. High EER ratings require larger evaporator and condenser coils. Additionally, fan motors may be larger.

A more subtle difference will be found in the two main electrical components—the fan motor and compressor. Units that were produced many years ago used low-efficiency fan motors which demanded high operating current—sometimes as much as that of today's modern compressor. Virtually all the energy spent in operating the fan turns into heat and none of it produces cooling. Obviously, to increase the EER rating of a unit, the manufacturer must use a more costly motor which has more iron and copper in its construction. Additionally, the old split-phase motor has given way to the more efficient permanent split-capacitor motor. This requires an extra capacitor in the unit.

The compressors used in modern air conditioning systems have also undergone an efficiency evolution. They are smaller, lighter, more efficient, and probably less costly to manufacture (taking inflation into consideration). However, one does not get something for nothing: The modern compressor of today is harder working, but it may have a higher failure rate than the old workhorses of the past.

3

Air Conditioning Systems and Equipment

3.1 General Information

Residential and commercial air conditioning systems have been around for many years and it is not surprising that they have evolved from the heavy, bulky, low-efficiency units of 20 to 50 years ago to today's smaller, lighter, and more efficient units. Many air conditioning systems were so well designed in the past that it is not unusual for the A/C service technician to be called upon to repair a unit that was produced 25 or more years before, and still is serviceable.

Fortunately, many air conditioning units are designed around the same principles of operation and use similar components. This permits the technician to replace older parts with new designs, often with only a minor modification for a different mounting or connection. Using a new design compressor or motor, for example, can result in an improved efficiency rating.

3.2 Compression Systems

Most residential and commercial air conditioners in service today operate with a mechanical compressor and use a refrigerant called Freon 22, which is sometimes referred to as R-22. Although, beginning in 1990, some Freon refrigerants were being subjected to a federal tax due to their ozone depletion potential, R-22 was exempt in that year. However, the tax laws may be changed in future years to include R-22 to help curtail its use. The federal tax is based upon the propensity of a CFC (chlorofluorocarbon) refrigerant to destroy the ozone layer in the atmosphere.

The service technician needs to be fully informed about replacement refrigerants which are presently being developed to take the place of the Freons, which have been standard for many decades. In the ensuing years, many new refrigerants will be used in place of R-12 and R-22. Some of these will also be eventually phased out of production as newer, permanent refrigerants are developed.

Some residential or commercial air conditioning systems employing R-12 (Freon 12) are still in operation. These are very old designs, and the use of R-12 for such air conditioning service was discontinued many years ago in favor of the more efficient R-22. R-12 and other refrigerants may be used in air conditioning systems which are designed for special purposes.

R-12 (or a suitable replacement) is required to be used in automotive air conditioning systems that contain rubber components, such as hoses, to avoid the problem of deterioration. The service technician should never recharge a unit with a refrigerant that is not identical to the original unless the substitute is recommended by the manufacturer of the equipment.

As the production of CFCs is reduced and eventually eliminated through strict environmental laws worldwide, there will be substitute refrigerants available which can take the place of R-12 and R-22. However, it is important to follow the manufacturer's directions for any alternative refrigerant so that the performance of the unit will not suffer. It is possible in some instances that modifications to the equipment, such as changes in the capillary tube or expansion valve, may be required when charging the system with a refrigerant other than the original.

3.3 Absorption Systems

Not all air conditioning systems use the refrigerant compression cycle as the method by which cooling by refrigeration may be accomplished. The ammonia absorption system, which has been used for residential refrigeration and commercial air conditioning systems for many years, is a viable alternative to Freon compression systems. It is becoming more popular since such systems offer certain economic advantages over compression cycle units.

The biggest advantage of the absorption system of air conditioning is its lower operating cost as compared to the more conventional

Freon compression cycle. An added advantage is a possibly significantly lower noise level. However, one does not get something for nothing; absorption air conditioning systems are more complicated, have higher original cost, and generally will require significantly more servicing during the life of the equipment.

Since the absorption air conditioning system operates on heat rather than mechanical energy, it may be favored in newer designs which allow solar heat to be used as part of the source of energy. As the cost of electrical power rises, air conditioning systems powered by the sun will become more and more commonplace.

With the increased popularity of recreation vehicles, the absorption cycle of refrigeration for this application is becoming widespread. Since this system of refrigeration uses heat, not mechanical means, as its source of energy, it becomes a practical matter to operate the vehicular refrigerator from propane, gasoline, the 12-volt vehicular electrical system, or 120 volts AC from a 60 Hertz power line.

The ammonia absorption system, being very different from compression systems, requires special equipment and techniques for maintenance, fault diagnosis, and repair. The service technician must be knowledgeable about such systems to properly service these units. Absorption system air conditioning service and repair is covered in Chapter 10 of this book, and the information presented in that section can also be helpful in the repair of domestic absorption refrigeration systems.

3.4 Heat Pumps

Another method to implement interior temperature control is through the use of a specially designed system commonly referred to as a heat pump, which is essentially a central air conditioner that can transfer heat in two directions according to the requirements of the season.

During the cooling season, when it is desired to transfer interior heat to the outside, the heat pump operates as an ordinary air conditioner. In winter, the function of the evaporator and condenser is reversed so that heat is transferred from the outside air to the controlled area.

Although the temperature of outside air during the heating season is relatively low, it contains a substantial amount of heat which can be recovered by a refrigerant compression cycle. Heat pumps are very popular in milder sections of the country where the outside temperature rarely goes below freezing, and are now becoming an alternative to other forms of heating and air conditioning everywhere.

A heat pump is more efficient than electric heat; for every kilowatt of electrical energy it consumes, it can transfer more than 1 kilowatt (3413 BTU) of heat energy from winter outside air to inside a building. Efficient heat pump design can result in an operating efficiency three or four times that of electrical heat. However, as outside temperature falls, heat pump efficiency also decreases.

3.5 Residential Window Air Conditioners

3.5.1 Construction

Window air conditioners are the most popular type of air conditioning units since they are relatively low in cost, small in size, and easily installed in almost any location. They are probably the simplest type of air conditioning system that the service technician will ever encounter. They are necessarily designed with relatively few parts, and diagnosing and correcting a problem is not complicated at all.

Figure 3.1   Photo of Typical Window A/C Unit

A typical small window air conditioner (Figures 3.1 and 3.2) will consist of a compressor, condenser, evaporator, and fan motor at the very least. It almost always employs a capillary tube as the refrigerant restricting device. Minimum electrical accessories will include a start and run capacitor for the compressor and a power on-off switch. Better quality units will also employ a permanent split capacitor (PSC) fan motor, and a thermostat.

Units which are designed for the lowest possible selling price (discount store specials) contain no thermostat, and the power switch controls both the compressor and fan motor, simultaneously. The compressor runs constantly as long as power is applied to the unit. Often, the fan motor is single speed

Higher priced units will have a thermostat and multispeed motor. The thermostat usually controls only the compressor in accordance with inside ambient air temperature, but some units manufactured today

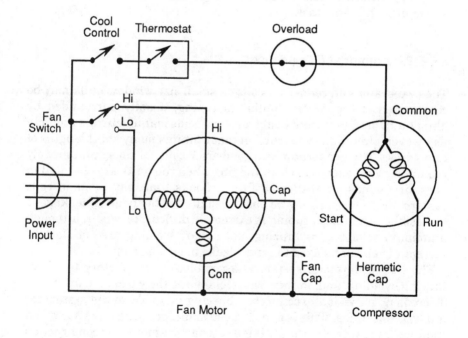

Figure 3.2 Typical Schematic Diagram
of Window A/C Unit

have an "energy saver" power switch setting, which allows the thermostat to control the fan motor in addition to the compressor. Some thermostats contain dual contacts, which allow the fan to cycle on and off independently of the compressor. A multi-function power switch allows the user to select automatic operation (thermostat-controlled fan for power savings), or standard operation, in which the fan runs continuously while the compressor cycles on and off.

Window air conditioner units as depicted above are manufactured in sizes as small as 4000 BTU and as large as 30,000 BTU or more. The larger size units, usually above 12,000 BTU, are powered by a 208/230 AC power line, almost always are constructed with capacitor run fan motors, and contain thermostats.

With the increased use of electronics technology in all types of electrical equipment, many manufacturers are utilizing solid state circuitry and electromechanical relays to provide the control functions in their air conditioning units. These parts replace the mechanical electrical components such as the multiple-position power switch and thermostat, and accomplish the same task.

3.5.2 Evaporator and Condenser

The evaporator and condenser coils on small size window units may be constructed of copper or aluminium tubing, usually surrounded by aluminum fins for efficient heat transfer. Some manufacturers prefer to use woven aluminum construction that contains many short lengths of fine wires. The evaporator and condenser coils on these units rarely require service, except to clean the fins when the units are operated in very dirty areas, or without a filter. Leaks that may occur in coils constructed of copper may be easily repaired if the point of leakage is accessible. Aluminum tubing is extremely difficult to repair, although aluminum soldering or brazing and heliarc welding are satisfactory methods that can be used to repair leaks in such tubing.

The sealed hermetic system is assembled at the factory by silver brazing, welding, and, in some cases, clamping the various connections. There is no provision to connect a charging manifold to the system for analysis or service. This is done to keep manufacturing costs down and improve reliability of the unit. It is a satisfactory method of construction since the sealed system of most window air conditioners is never serviced during its lifetime. A wide choice of clamp-on piercing valves for charging is available. Once so connected, it becomes a permanent part of the system.

3.5.3 Capillary Tube

Small size window units have one capillary tube connected between the bottom of the condenser coil assembly and the evaporator. This tube, sometimes coiled, may or may not be positioned so that it lies in the pool of water which remains in the bottom of the cabinet assembly. Some units are manufactured with the capillary tube soldered or mechanically fastened to the evaporator return pipe. These techniques are used by the manufacturer to provide a small measure of extra cooling of the liquid refrigerant in the capillary tube so that any gas bubbles contained in the liquid are condensed. This helps prevent the phenomenon of flash gas, which can be produced instantly by the liquid refrigerant as it passes out of the capillary tube.

Larger sizes of window air conditioners may have two or more capillary tubes connected between the condenser and evaporator. In order to transfer a greater quantity of liquid refrigerant in a given time, two or more smaller diameter tubes are easier to fabricate, and provide better refrigerant control, than one large diameter tube.

When the air conditioner is in the off cycle, the pressure built up on the high side of the system slowly dissipates through the capillary tube to the low side. This pressure equalization is a very important factor in air conditioning design because many compressors have such low starting torque capability that they cannot start with any appreciable unbalance in pressure between the suction and discharge ports.

3.5.4 Compressor

A wide variety of compressors can be found in window air conditioners. Some of these are the reciprocating type and others the rotary type. A new design, the scroll compressor developed by Copeland, will also find use in some window-type air conditioning units. The service technician should always attempt to use an exact equivalent compressor, when replacement is necessary, to restore the air conditioning unit back to its original specifications. However, in many instances, it may not be possible to obtain a replacement from the authorized parts supplier. In this case it is permissible to use a replacement unit of another design or manufacturer, providing that the electrical, mechanical, and BTU specifications of the new compressor are carefully checked against the requirements of the system. Other factors, such as starting torque and back pressure, must also be considered. Table 3.1 is a typical representation of replacement compressors used for air conditioning service.

The nameplate current rating of an air conditioner includes both the compressor and fan motor current. The full-load current rating of the compressor, as illustrated, will be less than the unit nameplate rating.

The replacement compressor manufacturer specifies the maximum BTU capacity obtainable from any given unit. The cooling capacity of an air conditioning unit may be 10% or more less than the maximum capability of the compressor, since other factors such as evaporator and condenser size, and air flow, will affect total cooling capacity.

Most compressors have two refrigerant connections—the suction or inlet tube, and the discharge or high-pressure connection. A third connection, called a process tube, is used at the factory for charging the system with refrigerant and is permanently sealed. In some units, in addition to the two refrigerant connections, there will be two more. These are used for cooling the refrigeration oil which is present in the crankcase of the system. Coolant is circulated through dedicated tubing which is part of the condenser assembly, and is cooled by the flow of air generated by the condenser fan blade.

Table 3.1   Table of Typical Replacement Rotary Compressors
for Single Phase Service

BTU Capacity	Voltage Rating	Full Load AMPS
6500	115	6.5
7700	115	7.3
9400	115	9.
11000	115	10.6
11000	208/230	5.9/5.3
15000	208/230	8.0/7.2
19000	208/230	10.0/9.1
24000	208/230	12.8/11.6
30000	208/230	16.6/15.0
36000	208/230	19.9/18.0
46000	208/230	25.5/23.0
60000	208/230	32.1/29.0

Hermetic compressors used in residential air conditioning systems are powered with either 115-volt or 208/230-volt single phase AC power. They employ capacitor run induction motors as part of the assembly. Some manufacturers provide a "hard start" circuit which utilizes a relay or solid state circuit to add additional capacitance or lower impedance to the circuit when starting. This increases compressor starting torque over the basic capacitor run circuit. Additional torque may be necessary in some units which are required to cycle on before the pressures on the low and high side of the compressor have equalized.

3.5.5 Overload Device

Compressors are always equipped with an overload device which monitors two parameters: compressor current and case temperature. The overload is a simple device, containing a bimetallic strip of material and a set of contacts, which are connected in series with the common lead of the compressor. Current flows through the material, which heats up in accordance with Ohm's law (power = resistance times current squared). In the event of a stalled compressor, the huge locked rotor current (10 times normal running current) provides an abnormal heat buildup in the metal. This results in it bending in such a fashion as to break the electrical circuit, protecting the compressor. Once tripped, the overload resets itself automatically when it cools. The overload control is connected in series with the common lead of the compressor as illustrated in Figure 3.3.

In most window air conditioning units the compressor is equipped with an external overload component which is field replaceable. Larger units, such as 18,000 BTU or higher, may have compressors that contain internal overload relays which are not field serviceable.

An overload protection device is always required in an air conditioning compressor circuit because there will be many times when the compressor will not be able to start due to head pressure, electrical, or mechanical malfunction. For example, an uninformed user may attempt to switch the compressor on before the required time has elapsed for system head pressure to dissipate through the capillary tube. A thermostat with too narrow a differential can result in "short cycling," which is a locked rotor condition. A momentary loss of power will also result in a stalled compressor, and the overload will protect the compressor from damage.

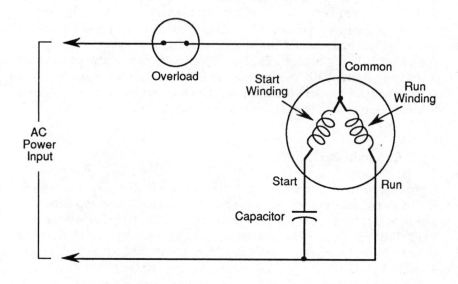

**Figure 3.3 Schematic of Compressor Circuit
Showing External Overload**

The compressor overload relay (Figure 3.4) has one additional function. An external unit is mounted in close contact to the compressor shell
so that its temperature is always very near that of the housing. During
sustained operation of the air conditioner, the temperature of the
compressor increases substantially. This is normal. However, under
abnormal operation, such as might be caused by a brownout or by
restricted condenser air flow, the temperature of the compressor will
increase above its design limit. The overload senses a sustained increase
in compressor temperature and opens the circuit to protect the compressor. If the compressor was not protected in this way, its life would be
shortened considerably. Internal overloads also protect the compressor
against thermal overload and once activated, may require a substantial
amount of cooling to restore circuit operation.

Connector
Terminal

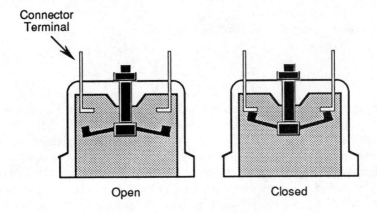

Open   Closed

Figure 3.4 Typical Compressor Overload
with Bimetallic Disc

3.5.6 Frost Sensor

Most air conditioning units are designed to be operated at outside air temperatures of 70 degrees F (21 degrees C) or more. Should a unit be operated at temperatures much below this level, the operating pressures within the system will be below normal. As a result, the liquid refrigerant flowing into the evaporator coils will vaporize at a temperature that is below the freezing level. This will cause the buildup of frost on the evaporator.

This condition is cumulative. The frost acts as an insulator and prevents heat transfer from the circulated inside air to the boiling refrigerant. As a result, more ice accumulates on the coils and eventually the entire evaporator can be coated with solid ice.

A point will come when the liquid refrigerant is not able to totally vaporize in the evaporator, and will return to the compressor where it can cause damage. Some manufacturers include an ice sensing thermostat, placed directly on the evaporator, which automatically opens the compressor circuit. This protects the compressor from damage.

This type of frost control is usually a single pole switch that is connected in series with the compressor common lead. Units which employ solid state control modules may use a thermistor in place of the switch. Electronic circuitry monitors the resistance of the thermistor (which varies with temperature) and inhibits compressor operation when the evaporator coil becomes too cold.

3.5.7 Capacitor

All compressors in window air conditioners utilize a capacitor connected to the start winding, which is permanently wired in the circuit. Although the term "start" is sometimes used to identify the winding of the compressor, the motor is actually a permanent split capacitor (PSC) design and the start winding is powered at all times when the compressor is running. The required starting torque for the compressor motor is provided by the electrical characteristic of the capacitor, in which the phase angle of the current leads the applied voltage by almost 90 degrees.

The capacitor usually has a rating of 15 or more microfarads, with a rated AC voltage equal to possibly twice the operating voltage, or more, of the air conditioner. This is a non-polarized, oil-filled capacitor, which may contain a vent to allow the escape of gas should it develop during abnormal operation. Some units are manufactured with built-in fuses, which prevent an explosion of the component in the event that it fails in a shorted condition. Some air conditioning circuits include a bleeder resistor across the capacitor terminals to dissipate any stored charge when the system is turned off.

Air conditioning units built years ago utilized capacitors which were filled with a compound called PCB (polychlorinated biphenyl), a carcinogen. Environmental laws have since precluded the use of this compound for capacitors used for A/C and refrigeration service, and systems built today use an environmentally safe product. One should be very careful in discarding capacitors which have an unknown filling compound. They should always be disposed of properly in accordance with local and federal environmental regulations.

3.5.8 Thermostat

Most window air conditioners are equipped with a thermostat that is mechanically operated either through the action of a bimetallic strip of material that bends with temperature changes, or a bellows actuated by a refrigerant-filled capillary tube that converts temperature to pressure.

With the advance of electronic technology, and specifically microprocessors, newer air conditioner designs which are now coming on the market utilize solid state circuitry to sense ambient room temperature and control compressor operation.

Thermostats are generally designed as single pole switches which are wired in series with the common lead of the compressor, so that they control its operation in accordance with the desired temperature setting selected by the user. Some thermostats are more complicated, using multiple switches which also can be used to control fan operation of the air conditioner if desired. Any unit which has an "energy saver" control on the front panel has this type of thermostat circuit.

Window air conditioner thermostats generally do not have narrow differentials, and may require a change in ambient temperature of several degrees F to effect switching action. One reason for using a relatively wide differential is to ensure that the head pressure developed by the compressor has had time to dissipate through the capillary tube during the off cycle. This allows the system pressures to equalize before the thermostat calls for cooling again. Should the thermostat have too narrow a differential, it can switch on before the compressor can restart. This is called short cycling and is symptomatic of a poorly designed system.

3.5.9 Bimetallic Strip Type

The bimetallic strip type of thermostat design (Figure 3.5) is the lowest cost thermostat, and is usually constructed with a microswitch to control compressor current. A microswitch is a spring return push-button-type switch that requires an extremely short linear motion for activation.

This type of thermostat can be designed either of two ways: The switch can be normally off and a rise in ambient temperature causes the bimetallic strip to bend in the direction to press the plunger and actuate the switch. Conversely, the switch can be designed to be normally on, and a fall in temperature causes the strip to bend and actuate the switch to shut off compressor operation.

The bimetallic-type thermostat is usually located behind the control panel of the air conditioner. After many months or years of operation, such thermostats can become loaded with dust and dirt which might inhibit proper operation. Any unit of this type which seems to exhibit intermittent compressor operation should be checked for a dirty thermostat and cleaned if necessary.

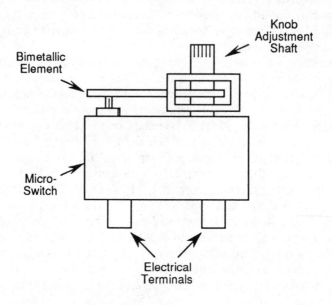

Figure 3.5   Bimetallic Thermostat

3.5.10 Capillary Tube Type

A more popular type of thermostat used in window air conditioners is the limited vapor-fill temperature control which employs a capillary tube, and sometimes a storage bulb, to sense temperature. This is illustrated in Figure 3.6. The refrigerant used in the temperature-sensing portion of the thermostat may be a refrigerant or some other suitable compound based on its ability to develop the required pressure to operate a bellows or similar device.

This type of thermostat has an advantage in that the temperature-sensing portion of the assembly can be placed any distance from the body of the unit using a length of capillary tubing. This allows the designer of the air conditioning unit to place the thermostatic control in the desired location without regard to its proximity to the evaporator, where room ambient temperature is usually sensed.

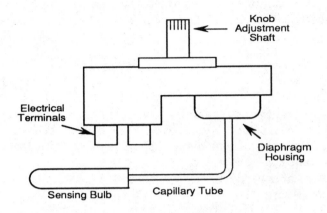

Figure 3.6   Capillary Controlled Thermostat

A limited vapor-fill thermostat has its capillary tube and sensing bulb (if so equipped) partially filled with its own self-contained refrigerant charge. Part of the assembly contains liquid refrigerant, and the remainder contains superheated vapor. This concept uses the pressure/temperature relationship of a saturated liquid/vapor mixture to generate a pressure within the tube. Temperature sensing occurs at the coldest part of the assembly, where the liquid refrigerant collects.

The thermostat is designed so that increasing ambient temperature, as sensed by the capillary tube or bulb, produces increased pressure which is used to provide mechanical displacement of a diaphragm or bellows. This, in turn, actuates a switch that is connected in series with the common lead of the compressor to control its operation.

The moving parts of this type of thermostat are usually enclosed in a housing, providing protection against dust and dirt. This greatly enhances its reliability over the bimetallic type which may have exposed moving parts.

Thermostats used in window-type air conditioning units are usually field adjustable for range, and sometimes for differential as well. These adjustments are usually accomplished by using a small screwdriver to turn the adjusting screw(s). Some thermostats are designed so that the adjustments are accessible only after removal of a cover plate.

Thermostats, and other operating controls, are usually serviced by removing a panel from the front of the unit. The compressor and fan motor capacitor may also be placed behind the panel. This type of design permits electrical service without the need to remove the casing of the unit.

3.5.11 Fan Motor

Window air conditioners generally have just one fan motor, but some units, such as casement types, are designed with two. When one motor is used, it usually has a double-ended shaft so that both evaporator and condenser fans can be operated.

Most fan motors used for this type of service depend upon the movement of air, generated by the condenser fan blade, to provide cooling of the motor windings. These types of designs are called "air over." Motors which do not require forced air cooling are referred to as totally enclosed non-ventilated (TENV) types. Many fan motors contain thermal cutout switches, which interrupt current to the windings in the event of an overheated condition. After sufficient cooling, the thermal cutout switch automatically resets itself.

Two types of fans are employed in window air conditioner units— squirrel cage (radial air flow) and propeller (axial air flow). The manufacturer of the unit may use the same type blade for both evaporator and condenser or may use one of each, depending upon the economics and requirements of the design.

Induction motors used in window air conditioners are single phase and fall into two general categories. These are the split phase and permanent split capacitor (PSC). The split phase motor is the lowest cost type, and will be found on many air conditioning units. The permanent split capacitor motor is a higher-quality unit which uses less energy (for improved EER air conditioner ratings), and has found favor in better quality units. In many cases, replacement motors supplied by the aftermarket will be permanent split capacitor motors.

The rpm of an induction motor is a function of the number of poles of the motor and the line frequency, as illustrated in Table 3.2. Window air conditioner units generally use either 4 pole or 6 pole motors, depending upon the desired operating speed.

The actual rpm of an induction motor is affected very slightly by its load, and must be somewhat less than synchronous speed due to a normal operating condition called slip. Synchronous speed is the rpm of the magnetic field generated by the stator winding, and is equal to 3600, 1800, and 1200 rpm in 2, 4, and 6 pole 60 Hertz motors respectively. Note

Table 3.2   Nominal Full Load Speed of
Typical Induction Motors

Motor Configuration	50 Hertz rpm	60 Hertz rpm
2 pole	2875	3450
4 pole	1438	1725
6 pole	875	1050

that motors designed for 60 Hertz service may not be used on 50 Hertz power lines unless so specified by the manufacturer.

It is mandatory that an unserviceable motor be replaced with another with the same number of poles. Fan rpm is an inverse function of the number of poles as indicated in Table 3.2. Should a 4 pole motor be substituted for a 6 pole unit, for example, the fan speed will be about 50% higher and the motor will draw excessive current. This will actuate the thermal cutout in motors so equipped, or otherwise burn out the winding. Should a 6 pole motor be substituted for a 4 pole unit, fan rpm will be too slow, and the air conditioner will not be able to deliver its rated BTU output.

Figure 3.7   4 Pole and 6 Pole Motor, Inside View

Fan motor rpm is sometimes indicated on the motor itself. If this information is not available, rpm can be determined by taking the motor apart and physically counting the number of poles. Each discrete coil of the stator represents one pole. Figure 3.7 shows a comparison between a 4 pole and 6 pole motor.

An induction motor, which is designed for single phase service, must have some means to provide starting, since one phase of current passing through a stator coil does not produce any torque to start rotation. This is provided by placing a second winding (sometimes called a start winding) on the stator, which produces a magnetic field that is offset from that produced by the main winding. Some motors employ a centrifugal switch that disconnects the starting circuit when the motor approaches rated rpm.

The net effect of having two windings, each with its respective magnetic fields at some angle to each other, is to produce a torque in the rotor. This starts the shaft rotating in the correct direction and it quickly comes up to speed.

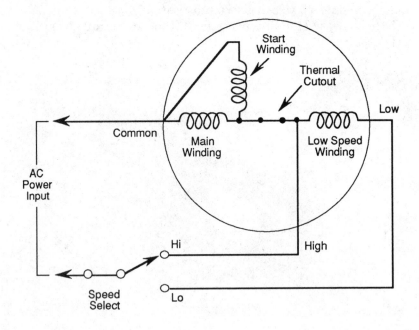

Figure 3.8   Wiring Diagram for Split Phase Induction Motor

A split phase motor (Figure 3.8) has a starting winding which is designed with more turns of finer wire than the running winding. The difference in inductance and resistance of this winding, compared to the main winding, causes a phase difference of the current in each winding. As a result, a second magnetic field is produced, which is offset from the main field.

A PSC motor (Figure 3.9) generates its phase shift in the current of the auxiliary winding through the action of a capacitor. This type of design more closely simulates a true 2 phase motor, in which the phases are separated by 90 electrical degrees. Phase shift in the second winding is produced by the capacitor, which draws a current that leads its applied voltage by almost 90 degrees. The capacitor winding remains energized as long as the motor is operated.

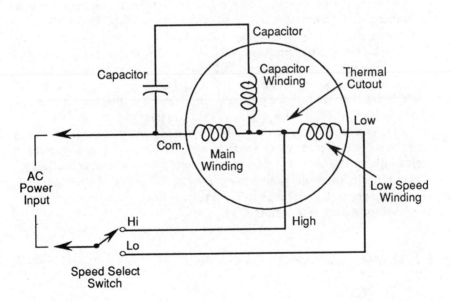

Figure 3.9   Wiring Diagram for PSC Induction Motor

3.5.12 Bearings

Motors used in window air conditioners may or may not be equipped with a provision to oil the bearings. Unfortunately, the trend nowadays is for manufacturers to use the lowest-cost motor available to them, and these components do not have quality bearings which can be serviced. Some motors are equipped with steel bearings, which do not have as long a service life as bronze. Such motors may become quite noisy after just a few seasons of use.

Fan motor bearing failure is quite common on low-cost, small-size window air conditioner units that see many hours of service. Bearings can fail in either of two ways, both attributable to lack of lubrication. They may wear out, resulting in excessive clearance between the shaft and inside of the bearing. This results in very noisy operation of the fan motor. Bearing replacement is theoretically possible in such a situation, but the service technician will generally replace the motor.

A second way that bearings can fail is for them to dry out and "freeze" so that the motor cannot come up to proper speed or rotate at all. This type of bearing failure is not always fatal. On motors that have provision for lubrication, the bearings can be oiled so that motor performance is restored. Other motors which do not have lubrication ports can also be oiled, but this procedure is more difficult and is not always successful. Such motors can usually be put back into service if they are completely disassembled and fresh oil is placed into the bearing assemblies.

Motors can also fail by developing shorted windings. This is generally a non-repairable condition. When a window air conditioner has a fan motor failure, the cost of parts and labor to replace the motor is often too great to justify the repair. However, if the problem is with the bearings, an astute service technician can often return the unit to service without the need for motor replacement.

3.6 Residential Central Air Conditioners

3.6.1 Description

Central air conditioners used in residential service are the "big brother" of window units since they have much in common with them. These units are split into two sections: compressor/condenser unit and evaporator/blower unit. A very large percentage of residential central air conditioners are used in conjunction with a forced warm air heating system, either as original equipment or as a retrofit, and use the furnace blower for circulation of the cooled air throughout the home. Such systems simply have an evaporator coil assembly installed in the furnace plenum and the compressor unit mounted nearby at the outside of the building.

Units designed for installations without the availability of warm air heating, such as in homes heated by baseboard hot water, use a self-contained air handler. This is a blower assembly and evaporator assembled into a cabinet, which is installed above the controlled area, usually in an attic. The compressor/condenser section is mounted outside on the ground or roof and a pair of refrigerant lines is used to make the connection between the two sections, just as in the central air conditioning unit that services a warm air furnace.

3.6.2 Power Source

Although the small window air conditioner and central unit have much in common, there are some differences which are illustrated in the typical schematic diagram of the central unit designed for operation with a warm air furnace.

Residential central air conditioners are usually powered by a 230-volt single phase power line (Figure 3.10), since the cooling capacity is generally 18,000 BTU (1 1/2 tons) or more. Many units are capable of proper operation over a range of 208 to 240 volts, which will be indicated on the nameplate.

The cooling thermostat circuit is usually operated at 24 volts AC, which is supplied by a small step down control transformer mounted in the compressor/condenser unit. When a single thermostat is used for both heating and cooling, isolation provided by the separate cooling low voltage transformer allows both heating and cooling thermostat circuits to be combined with no adverse effect. The 24-volt power is used to actuate the coil of a heavy duty contactor that controls line power to the compressor and condenser fan motor located in the outside unit. The furnace blower is controlled by a secondary relay, installed in the furnace section, that is energized by the thermostat circuit when cooling mode is in operation.

High-pressure and low-pressure limit controls are often used to protect the compressor against damage in the event of a malfunction that results in abnormal operating pressures. These controls contain normally closed single pole switches, which are connected in series with the coil of the contactor and prevent its operation should the system pressures fall out of normal operating range. The high pressure cutout is usually manually reset; the low-pressure switch may be automatically reset.

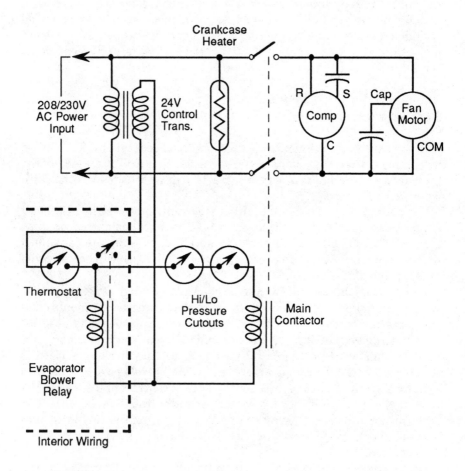

Figure 3.10   Wiring Diagram of a Typical Central A/C System

3.6.3 Evaporator

The evaporator coil used in residential air conditioning systems is similar to that of a window air conditioner unit, except it contains a greater quantity of tubing. The shape of the evaporator assembly can be different for various types of installations.

Many central air conditioning evaporators used in conjunction with a forced warm air furnace are shaped in the form of an "A" (called an A coil), and placed in the furnace plenum above the heat exchanger. Some evaporator designs for furnace service may be flat and installed at an angle above the heat exchanger. These are called slope coils.

The metering device that controls liquid refrigerant flow into the evaporator may be either a group of capillary tubes or an expansion valve. Since central air conditioning systems must transfer large amounts of heat, they may use four or more capillary tubes to deliver the liquid refrigerant to four sections of the evaporator. Evaporator pressure in such systems is a function of the total refrigerant charge, and is designed to be high enough to prevent freeze-up of the condensate formed on the cooling fins. For example, if R-22 is the system refrigerant, the design of the air conditioning unit is such that the evaporator pressure is not much less than 60 PSI under any condition of normal operation.

Some manufacturers have chosen to use an expansion valve to control the flow of refrigerant. The sensing element of the expansion valve is securely clamped to the evaporator outlet pipe so that it can monitor the warmest part of the assembly and cause the expansion valve to deliver sufficient refrigerant as required by the heat load of the system. The temperature of the evaporator outlet (superheat) is controlled by the expansion valve.

Systems that employ expansion valves can be very sensitive to the heat load placed upon the evaporator. For example, if the filtering system of the blower becomes clogged so that insufficient air flow results, the resulting closing of the expansion valve due to reduced evaporator temperature can result in lower than normal evaporator pressure. This will cause the evaporator coils to freeze, and the eventual loss of all cooling. In an extreme case, a low-pressure cutout could be activated, shutting down the system.

3.6.4 Evaporator Blower

The blower system for the evaporator section will usually be relay controlled in a large system. Many warm air furnaces, operating in the heating mode, employ thermostatic switches to operate the blower only when the furnace plenum has reached a preset temperature. When air conditioning is added to the system, the set of blower relay contacts is connected across the thermostatic blower control so the fan runs continuously when the compressor is powered.

High-quality forced air heating systems that are equipped for air conditioning service will also have some means to automatically control blower speed in accordance with the heating or cooling season. During the heating season, the furnace blower speed should be set only as high as required to maintain continuous air circulation, which greatly enhances human comfort.

During the cooling season, the relatively slow blower speed used in heating mode will be too slow to properly load the evaporator coils with warm, humid, return air. To do this, maximum blower speed is required.

Modern warm air furnaces have automatic speed control circuitry for both heating and cooling. Others may require a manual changeover in accordance with the heating or cooling season. Many blower motors designed for central air conditioning service are capable of operating at four discrete speeds.

Self-contained split A/C systems which employ a separate air handler may have a method to adjust blower speed such as by multidiameter pulleys. Once the correct speed has been determined by the installer, it should never require readjustment.

3.6.5 Compressor/Condenser Assembly

The compressor unit, containing the condenser blower motor and fan blade, is contained in a cabinet which is usually installed on the ground next to the structure it services. Roof mounting is also possible with residential air conditioning systems if the building layout requires it.

Most of the electrical components are also located in the compressor/condenser unit. This includes the compressor and condenser fan motor run capacitors, low-voltage transformer, and relay contactor. Some systems are also equipped with high-pressure and low-pressure cutoff switches, which protect the compressor in the event of a malfunction of the system.

The relay contactor is a heavy-duty, 2 pole device which is able to withstand the high inrush current demanded by the compressor when starting. The power feed which comes into the unit is wired directly to the relay contacts. The remaining two connections of the relay feed the common and run connections of the compressor. The compressor start capacitor is connected between the run winding and start winding of the compressor to provide a current which is out of phase with the main winding current.

Power feed to the compressor unit is provided through a disconnect switch or box which is in full view of the service technician when working on the outside unit. This is a safety requirement which will allow the technician to disconnect power when required.

During the off season, or even during an extended shutdown, refrigerant will migrate from the interior evaporator assembly to the compressor and collect in the crankcase as a liquid. This is the result of the gaseous refrigerant condensing in the colder sections of the system due to the temperature differential between inside and outside. The liquid refrigerant will dilute the oil in the compressor crankcase and prevent proper lubrication of the parts when operation is resumed. This condition, called secondary slugging, can occur and may cause premature compressor failure.

All quality central air conditioning systems will prevent refrigerant migration by using some method of ensuring that compressor case temperature does not fall below that of any other system component. Three methods are employed.

Many compressors are equipped with a wrap-around resistance heater which is strapped below the oil level. The heater is energized as long as power is supplied to the system through the circuit breaker or fuse panel, even though the unit itself is not in operation. Some manufacturers will employ thermostats which energize the crankcase heater only when outside temperatures fall below a predetermined minimum level.

Some compressor manufacturers, such as Tecumseh Products Company, provide immersion heaters in certain models. These heaters are self regulating and energy efficient. These operate in the same manner as strap-on heaters.

A third method is employed by the original air conditioning system equipment manufacturer (OEM), in which the design of the electrical circuit is to keep the start winding of the compressor energized at all times. The small trickle current which flows through the capacitor and start winding is sufficient to maintain compressor case temperature.

Since the power factor of the start circuit is relatively low, the electrical cost of heating the compressor is small.

When power to a system has been disconnected for an extended period, such as during the off season, the compressor should not be started until the power has been turned on for at least 12, and preferably 24, hours. This will allow sufficient time for the crankcase heater to vaporize the refrigerant that has collected in the crankcase. In an emergency, when the system must be started up immediately, the compressor may be operated for one second only and allowed to rest for a minute. This may be repeated about four times. The jogging of the compressor in this manner will help reduce the possibility of the liquid refrigerant from causing compressor slugging.

3.6.6 Compressor Capacitor

Residential central air conditioning units which are powered by single phase AC voltage will employ a run capacitor which is connected in the same manner as in a window air conditioner. The capacitance value is usually larger. Some manufacturers also employ a "hard start" circuit in which the compressor capacitor is momentarily paralleled with another capacitor of much larger value, or a solid state component, to increase compressor starting torque. A start relay or solid state circuit is used to make the momentary connection to the auxiliary component during the starting period.

Using a hard start system can prove to be an advantage in a residential installation, since it is quite possible that the owner will sometimes manipulate the thermostat soon after the off cycle has begun and attempt to restart the compressor under head pressure which has not had sufficient time to equalize. Without the hard start circuit the compressor will not start, and the locked rotor current demanded by the compressor will cause the circuit breaker to actuate. This could be an annoyance to the user.

Modern central air conditioning systems are manufactured with solid state control circuits which are capable of preventing compressor start-up until sufficient time has passed for pressure equalization within the system. These automatic time delay control circuits prevent abuse of the compressor, which can be caused by either thermostat manipulation or a temporary loss of AC power to the system.

3.6.7 Condenser Fan Motor

The fan motor which is used in residential central air conditioning systems is usually a 208/230-volt single phase component that is powered only when the compressor is operating. This is accomplished by connecting the power input leads of the blower motor directly across the power connections to the compressor.

Blower motors used in central air conditioners are generally of higher quality construction compared to those used in window units. Many are designed with lubrication ports so that yearly preventive maintenance can be performed by the service technician or owner. The presence of lubricating oil in motor bearings is extremely important for long motor life. Any motor which is neglected will eventually fail before its time.

The blower motor will usually be designed as a permanent split capacitor (PSC) type. This type of construction is superior to the split phase induction motor that may be found on low-cost window air conditioners. PSC motors have higher starting torque, and draw less current from the line as compared to a split phase motor. This enhances system EER.

As with the compressor run capacitor, the capacitor used with a blower motor is a non-polarized oil filled component. Capacitors that were used in older designs were filled with a carcinogen called PCB, which must be disposed of properly when defective and replaced.

High-quality central air conditioning systems have a two-speed or continuously variable speed condenser fan motor. The purpose of using multi-speeds is to provide only as much condenser cooling as required. During periods of relatively cool outdoor temperatures, such as nighttime, the condenser does not require maximum air flow as it does during midday. This feature not only conserves energy, but provides a quieter operating system during much of its operating time.

3.6.8 Receiver/Drier

Many central air conditioning systems, but not all, employ a filter/drier that is connected in the gaseous refrigerant return refrigerant line, located near or inside the compressor housing assembly. Sometimes a liquid line receiver/drier, located between the condenser and evaporator, is also used. The purpose of this component is to ensure that any moisture which is contained within the hermetic system is absorbed by a desiccant placed inside the dryer. Moisture in an air conditioning system is a contaminate and can cause breakdown of the refrigerant or lubricating oil, producing harmful acids. Additionally, any significant amount of moisture which may be present in a system will form ice at any location where the operating pressure/temperature relationship is below freezing. Such ice can clog capillary tubes or expansion valves, and shut down the system.

The filtering action of these components also prevents small metal fragments or dirt from passing through the system where they may cause clogging or damage.

Often, an air conditioning system will employ a "sight glass" located in the liquid line to allow the service technician to visually confirm that liquid refrigerant, and no vapor, is present during operation. A continuous stream of bubbles indicates gaseous refrigerant in the line and a low refrigerant charge. The presence of pure liquid signifies that the refrigerant charge is sufficient. The appearance of bubbles when the system is first started, or when it is shut down, is not necessarily an indication of low refrigerant charge.

Many sight glass assemblies also feature a built-in chemical which is used to indicate the presence of moisture. The color of the chemical is green or blue when there is a safe minimum amount of water in the system. Should there be an excessive amount of moisture, the color will turn pink.

Service connections for central air conditioning systems are usually located external to the compressor housing near the drier and sight glass. These are usually 1/4-inch SAE connections (Schrader valves) to which the charging manifold gauge set may be connected. The valves are automatically opened when the hoses are connected. Some air conditioning systems employ service valves which do not have spring-loaded stems and must be turned on after connection of the gauge set hoses. For this purpose, it is best to use the proper tool to operate the valve, such as a 1/4-inch square socket or service tool designed for this application.

3.6.9 Accumulator

Some air conditioning system manufacturers include an accumulator in the system. This is a storage container which is placed in the low-pressure gas line between the evaporator outlet and compressor suction inlet. Its purpose is to trap any possible liquid refrigerant which did not vaporize in the evaporator. This condition, called liquid floodback, may be caused by evaporator fan failure, clogged air filters, or even refrigerant overcharging. If liquid refrigerant reaches the inlet port of the compressor, it may cause slugging. Such liquid can also collect under the oil in the crankcase, starving the compressor bearings of proper lubrication. These conditions may lead to premature failure of the compressor.

3.6.10 Protective Components

Excessive high head pressure can be caused by a restriction in the flow of air through the condenser fins. In an extreme case, failure of the condenser fan motor will quickly result in extremely high head pressure which can cause damage to the compressor or other parts of the system. A high-pressure cutoff, located in the high-pressure gas refrigerant line, will sense an abnormal condition and shut the system down. When this happens, the system cannot restart until a reset button is pressed.

Abnormal low pressure can be caused by a low refrigerant charge, failure of the evaporator blower motor, or a stuck expansion valve. The latter two conditions can quickly cause ice to form on the evaporator coils and suction line, and may result in liquid refrigerant reaching the suction inlet of the compressor. Should this occur, the compressor can suffer irreparable damage. A low-pressure cutoff may be installed in the suction line to sense an abnormal condition and shut the system down.

There is one additional protective device: the compressor overload which operates in a similar manner as in a window air conditioner unit. However, most large-size compressors as used in central air conditioning systems are designed with the overload located inside the hermetic housing. It is connected in series with the common lead of the compressor, and functions by monitoring compressor current and temperature. This type of overload is not field serviceable.

The overload is designed to interrupt compressor current in the event that it senses an overload condition. Such a situation may occur when the compressor does not start when power is applied, as might happen through manipulation of the thermostat by an uninformed user, or a momentary power failure. Usually, these conditions will cause the circuit breaker (or fuses) protecting the system to be tripped first, since it is a faster acting device. Should this not happen for any reason, the overload will break the circuit when it reaches its trip temperature.

The main function of the compressor overload is to monitor operation of the compressor by sensing its temperature rise. Under overload conditions, such as may occur with partially restricted condenser air flow or excessive evaporator heat load, compressor head pressure increases above the normal operating range. This causes increased current draw from the power line and higher than normal operating temperature. The compressor overload, designed with a long time lag, interrupts current under a sustained overload condition.

The long-time constant of the compressor overload is also evident once it trips out. It is completely automatic in operation and resets itself when the compressor has cooled down below the overload cut-in temperature. This may take 20 minutes or more, depending upon ambient and compressor temperature. There is no choice but to allow the compressor to cool down sufficiently before operation can be attempted and resumed.

3.7 Commercial Air Conditioning Systems

3.7.1 General Information

Commercial air conditioning systems are much like residential central systems except that they can have cooling capacities many times larger than will be found in a typical residential installation. These systems can be "packaged units" in which the entire system is enclosed in one cabinet, or they may be split systems with the compressor unit separate from the evaporator and air handler.

Most of the information already presented under residential central air conditioners will also apply to commercial packaged or split system units. Since commercial systems usually have greater cooling capacity or have unique installation requirements, the service technician should be aware of some of the special characteristics which he or she may encounter when servicing these units.

Relatively small commercial systems, five tons or less, are very similar in design as residential units. The may employ one compressor which is powered by a 208/240-volt single phase AC power line. When air conditioning systems are capable of 10 tons or more of cooling, they may employ two or more compressors which are controlled by a two-stage thermostat. This method of providing large cooling capacities has several advantages.

A two-stage thermostat actually contains two thermostatic switches in one assembly, with the set point of the switches offset by one or two degrees from each other. This arrangement allows one compressor (and its complete system including condenser and evaporator) to be activated first on a rise in temperature, providing half the total cooling capacity of the system. Should the heat load be too great for half-system capacity, the slight increase in ambient temperature above the stage 1 thermostat setting will cause the second compressor to kick in. When the controlled interior temperature falls below the set point of the second system, it reverts back to one operating compressor.

This type of arrangement has two important advantages: It is a more efficient system since only one compressor will be operating most of the time, using about half the total possible electrical energy. Additionally, it automatically provides a completely self-contained redundant system (except for possibly the air handler blower) which can take over in the event of failure of either the primary or secondary system. Failure of either part of the system allows only half-capacity operation.

Some commercial air conditioning systems employ water cooling of the condenser. This is accomplished by using a coolant, usually an antifreeze solution, to remove heat from the condenser coil in a heat exchanger. The solution is then pumped from the condenser heat exchanger to the roof (or other part) of the building where it is exposed to the atmosphere by spraying, and cooled by conduction and evaporation. The water is then recirculated back to the heat exchanger and the cycle is repeated. Fungus and bacteria inhibitors are often added to the coolant to prevent buildup of undesirable substances. Water-cooled air conditioning systems are usually equipped with protective devices which shut down system operation in the event of improper or total failure of the water cooling function.

3.7.2 Power Source

Commercial air conditioning systems may be powered by either single or three phase AC power, usually depending upon the cooling capacity. Very large installations will operate from higher than 208-volt three phase power. Units which are rated at five tons or less will often require single phase power; significantly larger units are usually operated from three phase power lines.

It might be noted that compressors (and motors) which are operated from three phase power lines do not require a run capacitor, as do single phase units. These types of motors inherently have high starting torque. The direction of rotation is automatically determined by the phase sequence of the AC power feeding the unit.

Blower motors for the condenser and evaporator, being relatively smaller HP units as compared to the compressor motor, may be operated from one of the phases (single phase operation) even though the system utilizes three phase power. It is entirely possible that the blower motors in any given installation powered by three phase current will also use three phase blower motors. Such motors and compressors are more efficient and can be manufactured in smaller configurations than single phase equivalents.

3.8 Heat Pumps

3.8.1 General Information

Cold winter air may seem to be devoid of heat, but it does contain sufficient energy which can be extracted by a properly designed heat pump. Although a traditional air conditioning system is technically a heat pump, this term is used for a specialized variation of A/C equipment in which the role of the evaporator and condenser can be interchanged by means of a system of reversing valves. Using this technique, heat can be transferred from a lower-temperature location to one which is higher, regardless of the season. Heat pumps described here refer to those which use the refrigerant compression cycle for transfer of heat.

Heat pumps (Figure 3.11) are popular in many parts of the country where the summer BTU cooling requirement is comparable to the winter BTU heating requirement. This allows the use of one system which will provide cooling or heating, as required. When used for heating purposes, the heat pump can be as much as four times as efficient as electric heat. For example, electric heat produces about 8500 BTU for a power input of 2500 watts of electrical power. A heat pump, using the same power, can deliver 30,000 or more BTU of heat, all extracted from the outside air. Heat pump efficiency, however, is best at outside temperatures above freezing, and falls as the temperature drops. When the heat pump can no longer supply sufficient heat energy to provide the desired comfort level, auxiliary fossil fuel heaters are activated.

When used as an air conditioning system for summer cooling, the heat pump operates as a typical compression cycle central air conditioning system. In cooler weather, when it is desired to provide interior heating, the path of the refrigerant is changed by a reversing valve so that the hot, compressed gas is passed through the interior coil (now operating as a condenser) where it gives up its latent heat as it changes to a liquid. The liquid refrigerant is then fed to the outside coil which extracts heat from the air as the refrigerant changes back to a gas.

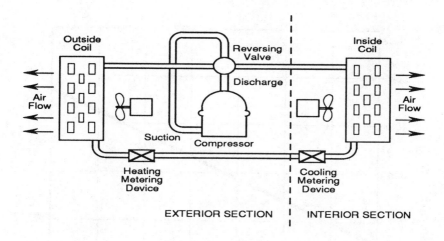

Figure 3.11   Basic Heat Pump Depiction

Even when the outside temperature is as low as 10 degrees F (–12°C) the liquid refrigerant can be boiled at a lower temperature, such as 0 degrees F (–18°C) to extract heat. However, operating an evaporator coil at temperatures below freezing causes ice formation on the coils. This problem must be addressed in the design of heat pumps, which includes a defrost cycle to melt accumulated ice.

Defrosting can be accomplished by the use of electrical heaters which kick in when needed. An alternate method is for the system to automatically revert back to cooling mode where the hot, compressed refrigerant gas is passed through the outside coil to melt the accumulated ice.

Heat pump efficiency (coefficient of performance, or COP) is a measure of BTU output as compared to the BTU input demanded by the unit (electrical input power in watts). COP varies inversely with the prevailing outside temperature. A typical three-ton unit (36,000 BTU/hr heating capability for example, may be able to deliver only 14,000 BTU/hr when the outside air temperature is 0 degrees F (–18 degrees C). Typical heat pump performance, as a function of outside air temperature, is illustrated in Figure 3.12.

For special applications, heat pumps can be designed with outside coils, which are placed below ground level, to extract heat from the earth. This can be a practical solution to providing interior heating at locations

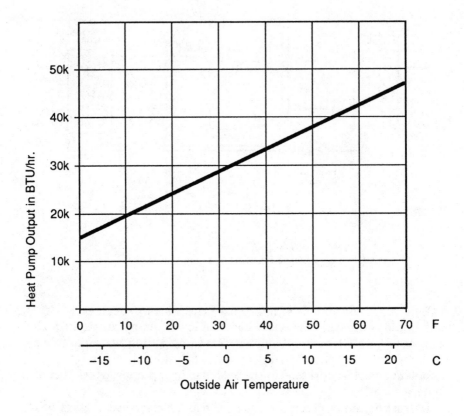

Figure 3.12 Heat Pump Performance Graph (Three Ton Unit)

where the air temperature is very low. Ground temperatures will remain fairly constant, and above the freezing level, during cold winter spells.

In order for a heat pump system to provide sufficient heat during extremely low outside temperatures, it is often designed with multistage electric heaters which are automatically activated as necessary. These heaters also come into play during the defrost cycle, when heat is removed from the interior coil. Many heat pump systems operate with existing conventional gas or oil heating systems which provide backup heat during very cold weather.

3.8.2 Reversing Valve

The main difference between a heat pump and an ordinary air conditioning system is the reversing valve, which controls the direction of refrigerant flow throughout the system. A schematic diagram illustrating both modes of operation is shown in Figures 3.13 and 3.14.

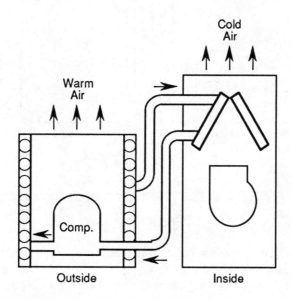

Figure 3.13   Heat Pump System Cooling Mode

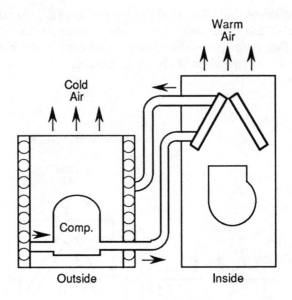

Figure 3.14   Heat Pump System in Heating Mode

Illustrated are simplified diagrams showing how the direction of refrigerant flow allows the heat pump to provide both heating and cooling as required. The switchover from heating to cooling, and vice versa, is provided by the reversing valve that is usually located in the outside section of the unit. There are many designs for reversing valves, and a system may contain two or more separate components. They are usually solenoid-operated so that they may be automatically activated by the demands placed on the system.

Since the heat pump system employs a separate metering device for heating and cooling, check valves are usually placed across the metering devices (expansion valves, capillary tubes, or orifices) to allow the free flow of refrigerant (in one direction only) to bypass the inactive restrictor. The direction of flow of the refrigerant automatically opens the check valve as necessary. When the check valve is closed, the metering device takes over control of refrigerant flow.

3.8.3 Defrost Cycle

Heat pumps must be capable of automatically initiating a defrost cycle when outside frost builds up on the outside coil. Some units will automatically switch over to "air conditioning" mode so that hot compressed gas is periodically fed to the outside coil to melt accumulated ice. The fan motor is inhibited during this cycle. Other designs will use electric heat to perform the same function. Ice must not be allowed to accumulate on the coils, since it impedes both air flow and heat exchange between the refrigerant and outside air.

Many heat pump systems manufactured today employ microprocessor controls which automatically determine when defrosting is necessary and initiate the defrost cycle. Such controls use information that is provided by various sensors such as thermistors, and allow the logic circuit to initiate and stop the defrost cycle as required.

3.9 Ammonia Absorption Systems

3.9.1 General Information

An alternative to the refrigerant compression air conditioning systems is the ammonia absorption cycle, which uses heat as its energy source rather than mechanical motion as produced by a compressor. The source of heat may be natural gas, propane (LPG), electrical, solar, or any other viable heat supply. One advantage of the absorption system is that the source of heat energy is not relevant; many units are designed to perform with more than one kind of heat source. An example of this would be a modern solar operated air conditioning system which uses natural gas as a backup when insufficient solar heat energy is available from the sun.

3.9.2 Absorption Cycle

Although absorption cooling systems operate on a somewhat different principle than compression systems, they employ similar components such as a condenser and evaporator, each of which function in the same way in both systems. Cooling is accomplished through the principle of changing the state of a refrigerant from liquid to gas to extract latent heat. The refrigerant that is used in many absorption systems is ammonia (R-717).

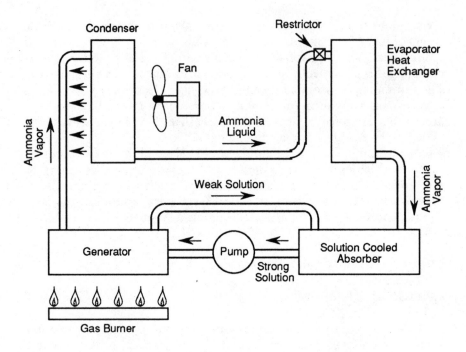

Figure 3.15   Simplified Ammonia Absorption System

The absorption system (Figure 3.15) uses an absorbent, usually water, which has the ability to quickly absorb large quantities of gaseous ammonia when cold and release it when heated. Ammonia vapor can be absorbed very quickly by cold water, just as a compressor can absorb gaseous refrigerant at its suction inlet.

The simplified system illustrated is a continuous absorption refrigeration cycle in which evaporator cooling is provided as long as heat is applied to the system. In addition to the evaporator and condenser, which perform the same function as in a compression cooling cycle, there are two other main components. These are called the generator and the absorber.

The cooling cycle begins in the generator, which contains a large quantity of a solution of water and ammonia. Heat provided by an external source forces the ammonia out of the solution in the form of a high-temperature, high-pressure gas which travels to the air cooled condenser where it gives up its latent heat and changes into a liquid.

The liquid ammonia is permitted to flow into the evaporator through a restrictor called an evaporator spray header, which is a metering device. The liquid ammonia absorbs latent heat in the low-pressure evaporator as it changes into a gas. The evaporator is surrounded by a circulating water system which gives up its heat to the evaporating ammonia, resulting in cooling of the water to a temperature of about 40 degrees F (4 degrees C). The cool water is circulated to an air handler located in the temperature-controlled area to pick up more heat as the cycle repeats.

The low-pressure ammonia gas then travels to the solution-cooled absorber which contains a large quantity of cool, weak ammonia solution. This solution absorbs the ammonia vapor as fast as it is produced in the evaporator, which sustains the required low pressure. The water-ammonia solution, at about 30% concentration, is recirculated back to the generator where it is heated, completing the cycle.

The efficiency of the simple absorption system described above can produce only about 18 BTU of cooling for every 100 BTU of input heat, an efficiency of only 18%. Modern-day absorption systems attain higher operating efficiencies by using additional components which increase the concentration of ammonia solution leaving the generator, and provide additional cooling of the ammonia vapor going to the condenser.

3.9.3 Heat Exchanger

Many absorption systems are self-contained ground- or roof-mounted units that are located near the air conditioned area they service. Cooling is accomplished by using an additional heat exchanger, called an air handler, located in the air conditioned area. The system operates by circulating an antifreeze solution between the evaporator section of the system and interior of the building. Antifreeze is required to prevent any possible freeze-up of the water during wintertime. Additional components, such as defoaming agents and fungus inhibitors, are placed in the circulating water system.

The absorption system of air conditioning uses a series of motors and pumps, in addition to the air handler components which produce the transfer of heat from the inside of the structure to the outside. As a result, this relatively complicated system (compared to a compression system) requires greater maintenance during its lifetime.

Air Conditioning Service
Equipment and Problem Analysis

4.1 General Information

The techniques used in servicing and repairing air conditioning units or systems are not much different than those required in the repair of any mechanical or electromechanical device. It is assumed that the service technician is knowledgeable and proficient in the use of ordinary hand tools such as screwdrivers, wrenches, and other common tools of all types. In addition, he or she should be familiar with AC power, electrical circuits and induction motors, and the use of simple electrical instruments such as voltmeters, ammeters, and ohmmeters. Air conditioning systems, while relatively simple devices, utilize a broad range of disciplines in their design and operation and require reasonable knowledge and skills for proper diagnosis, service, and repair.

Many areas of air conditioning repair are unique, and these will be addressed in this chapter. Along with special diagnostic and repair techniques, there are many dedicated tools and instruments which have been designed specifically for air conditioner service.

4.2 Service Equipment

4.2.1 Hand Tools

A well-equipped tool box will contain an assortment of screwdrivers for slotted and Phillips head screws, open-end and box wrenches, pliers of all types, and a complete socket set consisting of 1/4-, 3/8-, and 1/2-inch drives. It is important to include in any fully equipped tool box both English and metric-sized tools. As the changeover to the metric system is evolving in the United States, it is possible that any given unit will contain both English and metric hardware and components.

Other tools which are less common, but used frequently in air conditioning repair, are Allen and spline wrenches, nut drivers, and refrigeration ratchet wrenches with 3/16-, 7/32-, and 1/4-inch square openings. Metric refrigeration wrenches are also part of a well-equipped tool box. Electrical hand tools such as long-nose pliers, diagonal cutters, and wire strippers are important items.

Special long length Allen wrenches are required for access to set screws that are located on fan blade bushings. This particularly applies to window-type air conditioning units that contain squirrel cage blower wheels. The center bushings often employ set screws which secure the fan blade hub to the motor shaft, and the only way to loosen these screws is by means of an extra-long Allen wrench. Such tools can be purchased, or easily fabricated, by brazing a length of a cutoff Allen wrench to a hexagonal or square length of steel rod.

Rounding out the assortment of common hand tools is the hammer, various sizes of files, and adjustable wrenches. Other specialized A/C tools, such as fin combs, simplify difficult repair procedures.

It is often necessary to repair or replace refrigerant tubing and fittings. For this purpose a complete assortment of tubing tools is available, and should be part of any complete air conditioning tool set.

4.2.2 Refrigeration Tubing Service Tools

Tubing is used to carry the liquid and gaseous refrigerant between the various components of an air conditioning system, and the service technician must be familiar with all aspects of its repair and replacement. Many specialized tools are available for this purpose.

Most tubing that will be encountered in air conditioning work is soft copper refrigeration tubing, which is supplied in various diameters as illustrated in Table 4.1. Although the vast majority of tubing sizes today are measured in English units, it is also possible that metric-size tubing will become commonplace. Such sizes include 6, 8, 10, 12, 14, and 15 millimeters outside diameter. In refrigeration and air conditioning work, tubing is measured by its outside diameter, unlike the plumbing industry where inside diameter is used.

Table 4.1   Table of English Measure
Refrigeration Tubing Sizes

Tubing OD (inches)	Wall Thickness (inches)
1/8	0.030
3/16	0.030
1/4	0.030
5/16	0.032
3/8	0.032
1/2	0.032
5/8	0.035
3/4	0.035
7/8	0.045
1 1/8	0.050
1 3/8	0.055

Refrigeration tubing may be supplied in a sealed configuration that contains an inert gas such as nitrogen, which prevents oxidation of the inside of the tubing. Whenever a piece of tubing is cut from a roll, the remainder should be brazed shut to prevent any possible entry of dirt and to keep moisture out. Such contaminants can cause problems in an air conditioning system.

In addition to soft copper tubing, which can easily be bent as required for the application (using a tube bender), it is possible to encounter other tubing materials such as aluminum or steel.

There are several popular methods used to assemble tubing and fittings together in an air conditioning system. These are silver brazing (sometimes called silver soldering), flared connections, and compression fittings. O-ring fittings are used in vehicular air conditioning systems. Most residential and commercial air conditioning units are constructed with silver brazed connections and use little or no fittings, both as a cost-saving measure and for greater reliability.

Tubing should always be cut using a tube cutter (Figure 4.1). This method is superior to using a hacksaw, primarily because no chips are produced and no filing is necessary to remove burrs. Air conditioning systems must be kept scrupulously clean, and any metal fragments that remain inside a length of tubing can cause havoc in the system. After cutting a tube, a hand reamer (designed for tubing use) can be used to remove unwanted burrs.

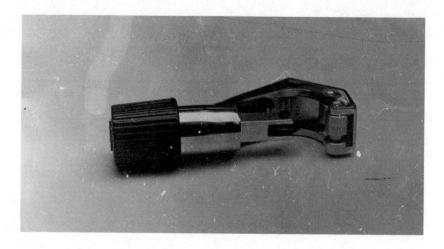

Figure 4.1   Tubing Cutter

Capillary tubing should never be cut with a tubing cutter since it tends to distort the copper and will restrict the small inner diameter. Capillary tubes may be cut by using a file to score the outside of the tube and bending it back and forth to break it. If necessary, any burrs may carefully be removed, taking care that no metal chips enter the tube opening, and that its diameter is not restricted.

4.3 Refrigeration Tubing Bending Procedures

Soft copper tubing can be easily bent by hand in a relatively large radius (compared to the diameter of the tube) if one is careful not to cause it to collapse at any point. For a more professional job, or where a tight radius is necessary, a tubing bending tool may be employed. The most simple type of tool is simply a tightly coiled steel spring which has an inside diameter equal to the outside diameter of the tubing. The same spring can also be used for bending a larger diameter tube by inserting the spring inside the part to be bent. Spring-type tubing benders are available in several sizes, and often sold in sets.

A more sophisticated type of tubing bender, which contains a wheel against which the tubing is bent, can also be used. However, for air conditioning service work this type of device is rarely a necessity.

When attempting to bend tubing, care must be taken to allow a bend radius of at least the outside diameter of the tubing. Smaller diameters are not practical, and can result in collapse and/or fatigue of the metal.

4.4 Tubing Connectors

Connections between tubing of equal or unequal diameters is accomplished by using brazed sleeve fittings, which are available in many sizes for that purpose. A far better method of connecting two copper tubes of equal diameter is done by swaging (Figure 4.2). This method does not require any fitting since the end of one tube is expanded so that its inner diameter is increased to the size of the outer diameter of the other tube. This allows a slip fit between the parts, and such a joint is more reliable since only one brazed connection is required.

A soft tube may be prepared for a swaged joint by using a specially designed punch which is hammered into the copper tubing to expand it to the desired size. Another method is by using a lever-type tool which expands the diameter of the tubing as the lever is squeezed. These tools are available to handle many different sizes of tubing, and either method prepares the tubing satisfactorily for assembly and brazing.

An alternate method to join two tubes of dissimilar diameters is to use a constrictor, which is a tool that reduces the diameter of the larger tube so that it is a reasonably good fit over the smaller tube. This is illustrated in Figure 4.3. Some tube-cutting tools are designed as a combination device, using one wheel for cutting and another for constricting. Since some silver brazing alloys are very tolerant of large gaps between parts, a constrictor is an excellent tool that can prepare two different sizes of tubing for a brazed connection.

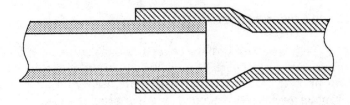

Figure 4.2 Swaged Joint Prior to Brazing Operation

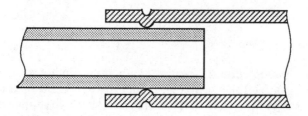

Figure 4.3   Constricted Joint Used for Tubing
of Different Sizes

It is also possible to join two soft copper tubes of different diameters without the need to use a swaged or constricted joint. This method is valid only if one tube may be slipped inside the other. It will work for a relatively large difference in diameters, and can successfully be used to braze a capillary tube to a larger soft copper tube.

To prepare the tubing for this type of joint as illustrated in Figure 4.4, both tubes are properly cleaned prior to assembly. The smaller tube should be inserted a distance of not less than 1/2 inch (1.2 cm). Capillary tubes should be inserted further, at least 3/4 inch (2 cm), to avoid any possibility of brazing material reaching the small capillary opening.

By using a Vise Grip® or arc-joint-type pliers, the larger diameter tube is squeezed so that its circumference hugs the smaller tube. The excess part of the larger tube is pinched together tightly into a U shape. Once the joint is mechanically sound, it may be silver brazed for a leak-tight connection.

4.5 Brazing/Soldering Torch

A brazing torch is required to make repairs and connections on tubing, evaporators, and condensers. A portable oxy-acetylene or air-acetylene torch will provide the high temperatures necessary for silver brazing. Some propane torches are designed for silver brazing work, but ordinary propane torches are not hot enough.

A very satisfactory alternate to an oxy-acetylene or air-acetylene system is Mapp gas, which is a combination of propane and acetylene that is supplied in a small cylinder similar to that of a propane torch.

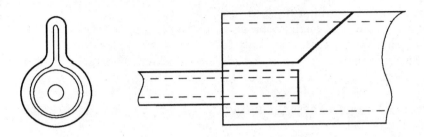

Figure 4.4   Mechanical Assembly of Two Different
Sized Tubing

The torch assembly comes equipped with a built-in regulator and it will develop the required high temperatures for silver brazing.

4.6 Silver Brazing

Silver brazing, sometimes called silver soldering, is the preferred method to join air conditioning tubing and components. It is performed at temperatures above 1000 degrees F (538 degrees C). Such connections, when properly made, are extremely strong and resistant to failures due to the ever-present vibrations to which they are almost always subjected. An added advantage is availability of silver brazing alloys which are designed to have tolerance to poorly fitted parts, which can be joined together for a leak-tight connection.

Silver brazing alloys are available in many compositions, each with its specific advantages. These usually contain 30% to 50% silver, and are not low in cost. Alloys containing very small quantities of silver are also available at lower prices. If possible, avoid any alloy which contains cadmium, which can generate hazardous fumes when vaporized during the soldering procedure. The vendor of silver brazing supplies should be able to recommend suitable alloys and fluxes for air conditioning repair work. A popular size for the brazing alloy is 1/16 inch diameter wire.

Alloys which contain appreciable amounts of silver will have a melting and flowing temperature in the 1100 to 1300 degree F (600 to 700 degrees C) range. It will be necessary to use a high-temperature torch, such as oxy-acetylene or Mapp gas. Ordinary propane torches do not have sufficient heat capacity for silver brazing, and should not be used.

In order to keep repair costs down, the service technician may use a soldering alloy, containing only 5% silver, at those locations where the high strength of silver brazing is not required. These alloys have significantly lower melting points than the higher-melting point brazing materials discussed above, and are in the 450 degree F (232 degrees C) range. These products produce a strong joint and can be successfully used with propane torches.

Certain procedures are necessary when making silver brazed connections. It is extremely important that the parts to be joined be absolutely clean of any dirt, grease, oil, or contaminants of any kind. Fine emery cloth or steel wool can be used to clean metal surfaces shiny bright, always being very careful to prevent any particles from entering tubing or components.

The joint should be mechanically sound before attempting to braze the parts together. Silver brazing exhibit remarkable strength, but can fail in service if joints are not properly assembled.

High temperatures encountered during the brazing operation will cause the formation of oxides on the interior surface of the tubing or connection, unless steps are taken to preclude such formation. Any oxide that forms is a contaminant to the air conditioning system and must be avoided.

An inert atmosphere may be provided by a flow of nitrogen gas through the system during the brazing operation, and continued until the parts have cooled. The flow of nitrogen should be set so that it is at as low a pressure as practical, possibly only about 1 PSI or so, to ensure that no air may enter the system during brazing.

Be sure to follow manufacturer's directions with regard to the proper flux to be used. The flux, usually in a pasty liquid form, is placed on the joint prior to the brazing operation. Care should be taken to ensure that no flux will enter the interior of the brazed joint where it will contaminate the sealed system. The purpose of the flux is to prevent oxidation of the base metal and absorb any residual oxides left after the cleaning process. It also provides the service technician with an indication of the brazing temperature, as it melts during the process.

Heat is always applied to the parts to be joined, and never to the brazing filler material. The flux will melt first before the parts have

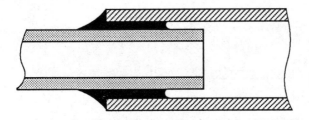

Figure 4.5   Brazed Joint Showing Smooth Fillet

attained the proper temperature for the brazing operation. As the filler material is placed against the joint between the two parts, it will melt when the correct temperature has been reached. Since the material will tend to flow towards the area of higher temperature, heat should be applied to the opposite side of the joint where the material is applied. Capillary action will draw in the filler as it is fed into the joint. When the connection appears full with a generous fillet of brazing (Figure 4.5), the operation is completed.

The joint may be allowed to cool naturally, or cooling may be hastened using a very damp cloth. Flux residue on the outside of the joint should be thoroughly cleaned off to prevent corrosion and to allow inspection of the connection. The appearance of the brazing material should be reasonably smooth, indicating a good connection. A leak test, to be performed later, will prove the validity of the joint.

4.7 Flared Connections

When replaceable components (such as driers) are part of an air conditioning system, methods of connecting the parts together other than silver brazing are sometimes used. A common practice is to use flared connections (Figure 4.6), which consist of specially prepared tubing and fittings which are designed to mate with the flare to produce a leak-tight connection.

Flares must be made with a special tool designed for the purpose. Flared fittings consist of two parts: the nut which fits over the tubing

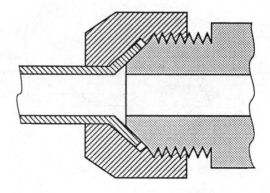

Figure 4.6   Flared Joint

Figure 4.7   Flaring Tool

and the male threaded piece against which the tubing will be held by the nut. Tubing must be properly flared to ensure a vapor-proof connection.

Flaring tools are simple devices which consist of a flaring block which is used to hold the tubing in place, and the cone-shaped flaring section
which forms the end of the tube to the correct 45-degree angle. Such tools are designed to flare an assortment of refrigeration tubing sizes. A typical flaring tool is illustrated in Figure 4.7.

Before attempting to flare a tube, the end should be cut off squarely and be free of all burrs. When using a file or emery cloth to remove burrs, any metal filings or dust should not be allowed to contaminate the inside of the tubing.

4.8 Compression Fittings

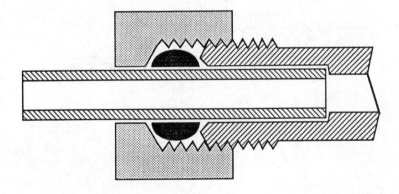

Figure 4.8   Compression Joint

Another method to couple tubing is with compression fittings, which are usually not used for refrigerant connections. They do find application, however, in water line connections. An offshoot of the compression fitting, called an O-ring fitting, does find extensive use in automotive air conditioning systems.

A compression fitting (Figure 4.8) is threaded assembly which makes use of an additional part, called a ferrule, to provide a leak-tight connection. The ferrule is placed on the tubing prior to its assembly into the fitting. When the nut is tightened, the compression of the ferrule against the tube and fitting makes the seal. A typical compression fitting will require one ferrule for each tube to be joined.

4.9 O-ring Fittings

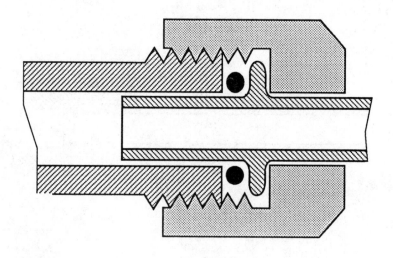

Figure 4.9 O-ring Fitting

O-ring fittings (Figure 4.9) employ a rubber ring, shaped like a doughnut, which provides the seal as the parts are assembled together. The seal is attained without the need to use a high amount of torque, as sometimes required with flared fittings. These fittings are found extensively in automotive air conditioning systems which employ connecting hoses between the various components of the system.

Prior to assembly of an O-ring fitting, the new ring (previously used rings should not be used) is coated with clean refrigeration oil. A backup wrench must always be used to prevent rotation of the parts as the connection is tightened. As the parts mate, the O-ring seals the connection with relatively little compression. The parts should be tightened only to the manufacturer torque specifications.

4.10 Process Tube Adapter

Connection to a hermetic system can be made by using a process tube adapter, in a similar manner as done during system charging at the factory. This method of access to the refrigerant system eliminates the need to permanently install an access valve, and results in a more reliable seal.

A process tube is a small length of tubing which is connected to the low-pressure side (and possibly also the high-pressure side) of the sealed refrigerant system. A process tube adapter may consist of a copper tube and valve assembly which can be attached to the low-pressure process tube to allow evacuation and recharge of the system.

When recharging is completed, the system process tube may be left connected to the system, properly capped, or pinched off. Pinching off is accomplished by using a tool (appropriately called a pinch-off tool) which mechanically closes the process tube to temporarily seal it. The valve assembly is then removed and the open end of the process tube is brazed shut, making a permanent seal. Once completed, the pinch-off tool is removed.

4.11 Capillary Tube Cleaner

A system with a clogged capillary tube may be encountered. This problem may be corrected by replacement of the tube, but an alternate method is to use a capillary tube chaser/cleaner.

This kit consists of a supply of lead alloy wires which are a few thousandths of an inch smaller than the ID of various capillary tubes. A short piece of wire, about 3/8 inch (1 cm), is inserted into one end of the capillary tube. A portable hydraulic unit, capable of generating up to 5000 PSI of pressure, is then used to force the wire through the capillary tube to clear the obstruction. The wire then falls into the evaporator to remain there harmlessly.

4.12 Manifold Gauge Set

One of the most important tools that the A/C service technician will need is the manifold gauge set (Figure 4.10), which consists of a pair of pressure gauges mounted on a dual-valve assembly. Three charging hoses are connected to the manifold by means of 1/4-inch SAE threaded fittings, sometimes referred to as Schrader fittings. Each hose has a pin in one of its connectors, which automatically depresses the valve stem in the Schrader valve when the hose connector is hand-tightened to the valve.

The manifold gauge set permits measurement of air conditioner operating pressures as a means to analyze system faults, to purge and recharge a system, or to monitor the operating pressure during recharging. The manifold gauge set contains a high-pressure gauge, and a compound low-pressure and vacuum gauge. A pair of hoses, one for each gauge, is connected between the manifold and unit under test to provide a reading of the existing pressures in the system.

The high-pressure gauge is capable of indicating pressures as high as 500 PSI, which is sufficient for measuring head pressure in any air conditioning system. The low-pressure gauge is a compound unit capable of measuring up to 80 or more PSI, and also up to 29.92 inches of mercury (a perfect vacuum at sea level). Some gauges also indicate vacuum in KPa. The low-pressure gauge has a retarded mechanism above its full scale calibration so that pressures as high as 250 PSI will not harm the gauge or pointer. The vacuum portion of the scale, usually calibrated in inches of mercury, is used when a system is being evacuated during a purging operation.

Both gauges contain a multitude of scales. In addition to the pressure and/or vacuum readings, the dial contains temperature readings which correspond to the evaporating or boiling temperature of various refrigerants. These calibrations permit determination of evaporator and condenser refrigerant boiling temperatures without the need to refer to a pressure/temperature chart.

Figure 4.10   Manifold Gauge Set
(Courtesy Robinair Division SPX Corporation)

Newer gauges also contain a metric pressure scale, kg/sq. cm, in addition to the conventional PSI scale. Additionally, some high-quality gauges are glycerine-filled to help mask rapid pressure variations which may be present in a system, allowing a more steady indication.

Two hand-operated valves are provided on the manifold gauge set. When the valves are closed, each gauge with its accompanying hose is completely separated and closed off from the center hose, and the gauges indicate the existing pressure at the access valve to which each is connected. Either or both valves may be opened to provide a path between the center hose and either or both outside hoses. The center hose is used to evacuate the system with a vacuum pump or to charge the system from a source of refrigerant.

The gauges and hoses are usually color-coded to help avoid misconceptions. For example, the low-pressure gauge and hose may be blue, the high-pressure side may be red, and the center hose may be yellow. Some manufacturers supply manifold gauge sets with four hoses, separating the function of evacuation and charging into two hose connections.

The hoses are constructed of reinforced rubber to withstand the high pressures encountered in an air conditioning system. They are terminated in 1/4-inch SAE threaded fittings with center depressor pins, to match the charging ports of most air conditioning systems. Rubber gaskets are placed inside the threaded fittings so that a gas seal is attained with finger pressure alone.

When an air conditioning system, such as a window unit, is not equipped with service valves, one must be added to the system before the manifold gauge set can be attached. Clamp-on piercing valves are used, and once installed left connected to the unit permanently.

Ordinary manifold gauge sets, which are constructed of brass and similar materials, should never be used when servicing an ammonia absorption air conditioning system. Ammonia refrigerant is corrosive to copper and brass. For such systems, steel manifold gauge sets must be used.

4.13 Vacuum Pump

The use of a vacuum pump (Figure 4.11) is required whenever an air conditioner has been opened to the atmosphere, or if it is desired to remove contaminants (such as moisture) which have collected in the system.

Figure 4.11   Vacuum Pump

Vacuum pumps are available as portable self-contained units of varying vacuum drawing capacities and displacement. Some units are capable of producing only 27 1/2 inches of vacuum (referred to as 29.92 inches of mercury as a perfect vacuum) and displacements of 1.7 cubic feet per minute. Higher quality units which are capable of drawing almost a perfect vacuum are rated in microns instead of inches. A micron is equal to one-millionth of a meter.

Vacuum measurements are often specified by the metric unit KPa instead of inches of mercury. Under standard conditions at sea level, atmospheric pressure is equal to 29.92 inches of mercury, or 101.3 KPa.

A vacuum pump with a capability of producing a vacuum of 10 microns or less will be able to remove more gas and moisture from an air conditioning system than one that can pump down to only 27 1/2 inches. Additionally, larger displacement pumps, such as 3 or 5 CFM, are able to evacuate a system faster than others with small displacement. The time difference may be important when a large commercial air conditioner is evacuated or when many units must be serviced in a short time. One method to decrease pumping time on a large unit is to use two vacuum pumps simultaneously.

Many vacuum pumps utilize a special oil to provide a seal as they draw a vacuum. This oil will become contaminated with use, resulting in increased operating time and less final vacuum. The oil should be periodically replaced to maintain pump performance.

Some air conditioning systems, such as heat pumps, require greater evacuation to a higher level than "cooling only" units. The low temperatures encountered in heat pump service cause any residual moisture in the system to have a greater propensity to form ice, resulting in a potential malfunction. Such systems should be pumped down to at least 500 microns when being evacuated. To measure such vacuum levels, vacuum gauges capable of indicating 50 microns or less are available.

4.14 Vacuum Gauge

The manifold gauge set contains a compound low-pressure gauge which is capable of indicating vacuum level, usually in inches of mercury. Such a gauge may not be sensitive or accurate enough when high levels of evacuation are required as in heat pump systems, which must be pumped down to levels of 500 microns or more (about 0.2 inches of mercury).

In order to properly measure such low levels of vacuum, an electronic vacuum gauge is used. These gauges are portable, battery-operated instruments which are capable of measuring vacuum levels as small as 50 microns or less, and provide the necessary accuracy for high-level evacuation procedures. Some gauges provide vacuum readout in both inches of mercury as well as microns.

4.15 Thermometer

The thermometer (Figure 4.12) is an important tool for air condi-tioner service. Measurement of the operating temperature of certain parts of the system will provide important data concerning a unit's performance. Analysis of the measurements can determine if a problem exists.

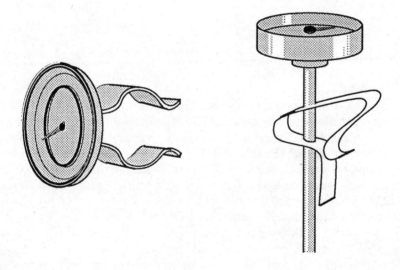

Figure 4.12   Clamp-on Thermometer

Almost any type of thermometer can be used to measure the air temperature differential across an evaporator or condenser. It is recommended that only those with 100% metal construction be used, since glass thermometers are subject to breakage in the harsh environment of air conditioning service and repair.

To aid in determining system performance, the temperature of liquid and/or gas refrigerant lines may be measured. For this purpose a clamp-on thermometer, designed specifically to measure the temperature of a tube or any other surface to which it is attached, may be used. Water-cooled systems often are equipped with thermometer wells that allow temperature readings using a straight shaft metal thermometer.

Measurement of the air discharge temperature of an air handler, or evaporator outlet air temperature in automotive air conditioning systems, is an important diagnostic function. Manufacturers supply charts of typical readings of discharge air temperature, as a function of outside air temperature and humidity, to help ascertain proper operation. Without this information, it may not be possible to prove that a system is operating normally.

4.16 Leak Detectors

All air conditioning systems will eventually lose part or all of their refrigerant charge due to leakage, since internal system pressure is greater than atmospheric pressure. Many air conditioning systems are manufactured sufficiently tight so that excessive refrigerant leakage is not a problem during the lifetime of the equipment. Leaks as small as one ounce or less per year, while not catastrophic in nature, can be located by sensitive detecting devices. For this purpose, several different methods of locating refrigerant leaks are available. The method employed often depends upon the magnitude of the leak.

To perform a leak test, the system needs to be somewhat charged. For large leaks, the pressure within the system can be as small as 5 or 10 PSI, and a soap solution (or commercially prepared leak detection product) will indicate bubbles at the point of the leak. The pressure within the system can be adjusted as required to produce bubbles at the suspected location of the leak.

Leaks that cannot be located by visual inspection or a soap solution will require more sophisticated methods. An excellent inexpensive leak detecting device is the halide torch leak detector (Figure 4.13). This is a propane torch system which has a burner head that is specifically designed for refrigerant leak detection. Halide leak detectors can locate refrigerant leaks of about eight ounces a year.

Figure 4.13   Halide Torch

The burner is designed so that its air supply must come through a flexible hose that is used as a sniffer by the service technician when searching out the leak. The purpose of this hose is to draw in, with the air required to burn the propane, any possible refrigerant gas that is emanating from a leak in the air conditioning system.

A copper plate placed inside the burner is heated red-hot by the propane flame. This acts as a reactor. The leak detection system makes use of the fact that a properly burning propane flame is almost colorless. Should a relatively small amount of refrigerant gas be present in the air supply, the resulting color of the flame will be green. Should a large amount of refrigerant enter the air-supply tube, the flame will turn bright blue.

The presence of a bright blue flame represents a hazard since the detector is producing phosgene gas, which is toxic. For this reason, the area should be well-ventilated, and if a blue flame is produced, the torch should be immediately removed from the source of the leak.

When a halide torch is used to locate leaks that are relatively large, it would be prudent to charge the system to a pressure only large enough to produce a green flame when the air hose is brought near the leak. Often, all that is required is a pressure as small as 5 or 10 PSI. When large amounts of refrigerant contaminate the atmosphere, the flame will assume a green color, making leak detection more difficult. Using portable fans to clear the area of refrigerant will help.

Very sophisticated electronic leak detectors are available at reasonable cost, about $150 or less. A typical electronic detector is illustrated in Figure 4.14. These instruments generally use the change in electrical conductivity of air, or a change in its dielectric constant, when it contains any of the commonly used refrigerant gases. They are extremely sensitive and can detect refrigerant leaks as small as 1/2 ounce a year.

Some units are designed to indicate the presence of refrigerant by producing a "geiger counter" sound, which is a low ticking rate. As the sensing tip of the detector is brought closer to the source of the leak, the rate increases dramatically and simulates a siren.

Electronic circuitry within the detector automatically compensates for ambient levels of refrigerant in the air, so that leak detection may be performed in a contaminated atmosphere. As with any type of leak detection method, only a minimum amount of pressure in the system should be used first, and increased if necessary. For detecting extremely small leaks, it may be necessary to pressurize the air conditioning system to a maximum of 150 PSI using nitrogen or other inert gas. This will force a greater quantity of refrigerant out of the leak, making detection easier.

Other types of refrigerant leak detectors are also available. These include the ultrasonic detector which responds to the high-frequency sound generated by a small gas leak, and the ultraviolet detector which visually pinpoints the source of a leak by detecting seepage of a special oil injected into the air conditioning system.

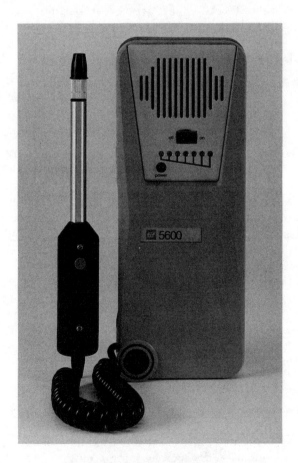

Figure 4.14   Electronic Detector
(Courtesy of TIF Instruments)

4.17 Inert Gas Supply

A supply of inert gas is part of a well-equipped service operation. Nitrogen is used to perform a purging operation, or to pressurize a system to an artificially high level for leak detection. Dry nitrogen is low in cost and available in steel cylinders. An alternate gas that may be used is carbon dioxide.

The pressure level in a cylinder of nitrogen is about 2000 PSI. Carbon dioxide pressure is 800 PSI. These are dangerously high pressures, and gas cylinders must be equipped with regulators to reduce the pressure to a safe level of 150 PSI or less. Additionally, cylinders must be supported by chains or other means to prevent them from being accidentally knocked over.

4.18 Refrigerant Recovery System

With the ever-increasing cost of refrigerants, and new environmental laws being enacted on their use and disposal, there may come a time to consider the investment of a refrigerant recovery system. This type of equipment is now available and can provide a method of recovering refrigerant that otherwise would be dissipated into the atmosphere. The cost of such systems can often be quickly paid back through the reuse of expensive refrigerants such as R-12.

Refrigerant recovery systems use a pump to remove the gaseous refrigerant from an air conditioning system, pressurizing it to a liquid so that it can be stored in a container. The used refrigerant is continually passed through filters and other components which remove contaminants, even those produced in a burned out system. The recycled refrigerant can then be used to recharge the system from which it was taken, or reused in any other application.

4.19 Manometer

A manometer is an instrument which is used for measuring relatively low pressure levels of gas or vapor, by using a column of water or mercury as the indicating mechanism. The pressure source pushes the liquid indicator, in a clear tube, against the normal force of gravity. This produces a reading of the pressure in inches of water, millimeters of mercury, or any other equivalent unit.

Manometers are used to measure air pressure drop across an evaporator or condenser coil, or the regulated natural or LPG gas pressure of an ammonia absorption gas air conditioning system.

4.20 Hermetic Analyzer

There are many hermetic analyzer instruments available to the service technician. Many of these units are capable of measuring current and voltage as well, and they are used to indicate the condition of a compressor. Some hermetic analyzers are also capable of providing a "reverse running" mode which can sometimes free a stuck compressor.

A hermetic analyzer is not a high-priority instrument. All of its functions can be simulated by means of manual test connections and performing electrical measurements using an ordinary multimeter.

4.21 Electrical Test Equipment

Air conditioning units are electromechanical devices, and the service technician must be familiar with elementary electrical concepts to properly diagnose and repair them. For this purpose, it is recommended that several electrical measuring instruments be part of a well-equipped tool box.

A wide selection of low-cost test instruments is available. In particular, battery-operated, portable, hand-held digital multimeters (Figure 4.15) can perform a multitude of functions. Such meters can be purchased from numerous supply outlets and can be obtained for less than $50. Dedicated meters which measure only one electrical parameter may also be used.

Multimeters measure voltage and current, both AC and DC, as well as resistance. Many are also designed to provide additional functions such as continuity checks and diode measurements. It is helpful to be familiar
with solid state devices such as diodes and transistors, since the trend in new equipment design is to use solid state controls in place of the older mechanical control components.

Figure 4.15   Small Hand-held Portable Multimeter

4.21.1 Voltmeter

The voltmeter function is used to verify that power is being applied to the various parts of the air conditioning system, such as blower motors, compressors, relay coils, and thermostats. For this purpose, it is necessary to measure AC voltages as low as 24 or less, and as high as 240 (or more for high-capacity commercial air conditioning systems). DC voltage measurements are not normally required for traditional AC systems, but the trend towards solid state air conditioning control devices makes such measurement capability mandatory.

A handy accessory to sometimes use in place of a voltmeter is a very low-cost pocket neon tester, which will indicate the presence of voltage. This can be used to quickly check for blown fuses, bad relay contacts, etc., where measurement of the magnitude of the voltage measurement is not required. Using a neon tester to detect the presence of voltage can save much time in diagnosing a problem. Neon testers are not used on voltages less than 100 volts.

4.21.2 Ammeter

Current measurements are made with an ammeter. In order to measure current with a traditional ammeter, one power line feeding the equipment must be opened to allow the meter to be connected in series with the circuit. The current to be measured thus flows through the instrument. Opening the circuit is not necessary if a clamp-on style meter is used. Such an instrument (Figure 4.16) contains a split loop of magnetic material which can be placed over a conductor to measure the current it is carrying. Most clamp-on ammeters have switchable ranges which cover current levels that will be encountered in air conditioning service and repair work.

Figure 4.16   Clamp-on Ammeter
(Courtesy of Amprobe)

An ammeter is one of the most important diagnostic tools for servicing window air conditioners. The current draw of the unit, compared to the nameplate specification, will quickly tell a great deal about the condition of the system. To facilitate such a measurement, a simple diagnostic ammeter can be built using a panel-type iron vane ammeter wired to a common AC plug and receptacle. The tester can be constructed using a heavy-duty, 14-gauge, three-wire line cord and plug, and the ammeter and receptacle may be mounted in a suitable enclosure. This tester is not voltage-dependent, and may be used on 115-, 208-, and 240-volt circuits
with the proper adapters. Figure 4.17 illustrates the wiring diagram of the diagnostic ammeter.

The tester is connected between the unit under test and power receptacle in a similar manner as an ordinary extension cord. It automatically connects the ammeter in series with one of the power lines feeding the air conditioner without the need to open any circuit. Ground circuit integrity is maintained through the test box. For dual voltage operation, adapters can be constructed to convert the 120-volt plug and receptacle to 240-volt types.

This handy service accessory will provide measurement of both fan motor and compressor current. To measure fan motor current, the unit can be set to fan only, or the thermostat can be set to prevent compressor operation. It will often be the first tool used in the diagnostic procedure, and can help save much valuable service time.

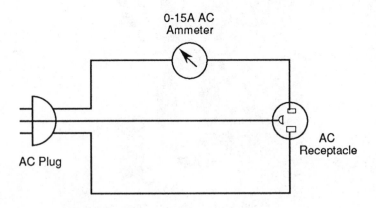

Figure 4.17   Wiring Diagram of Line Current Test Instrument

Some air conditioning systems require DC current measurement capability. In particular is the gas-operated ammonia absorption system, which contains an electronic ignition circuit. A DC microammeter is an instrument that is used to detect the extremely small operating current, which is in the range of just a few microamperes. One microampere is equal to one-millionth of an ampere.

4.21.3 Ohmmeter/Continuity Checker

The ohmmeter, usually part of a multipurpose digital or analog test instrument, is used to measure resistance of electrical components, as well as determining if a connection is valid or if it is shorted to anything else. When used for continuity checks, the ohmmeter is set to the lowest scale, usually R X 1. This permits verification of electrical connections which will indicate a resistance reading of much less than 1 ohm. The ohmmeter and continuity checker are always used with all power disconnected from the unit under test.

To measure leakage paths or motor winding resistance, higher resistance ranges, as necessary, may be used. For detecting electrical leakage paths, such as from component wiring to ground or unit frame, the highest resistance range of the instrument is used.

4.21.4 Megger

An ordinary ohmmeter uses a very low, safe voltage (possibly as small as 1.5 volts) for resistance measurements. This low voltage is not representative of actual operating conditions of compressors and induction motors, which see 150 or more peak volts. A megger is a special type of ohmmeter which utilizes high voltage, 500 volts or more, to perform a non-destructive test on electrical components. It is capable of measuring very high values of resistance. This procedure can reveal leakage problems which cannot be detected by an ordinary ohmmeter.

Use of the megger is especially valuable in determining the electrical condition of hermetic compressors. Although the internal wiring of these components is fully insulated and one would expect an infinite resistance reading between the windings and shell of the unit, this is often not the case.

Compressor windings are drenched in oil and refrigerant, and if either is contaminated, the winding resistance can be in a process of deterioration, resulting in an electrical leakage path to the case or shell. Should the leakage become great enough, the compressor will fail.

The megger is becoming popular as a diagnostic service instrument. The electrical condition of a compressor can be determined, and its remaining electrical life determined. If a problem is detected, remedial action can often be taken to eliminate or diminish the source of the problem. Moisture and contaminants can be removed by replacing compressor oil, purging, and evacuation. The installation of a filter/drier to the system can help remove moisture and particles from the system. After remedial action has been taken, higher megger readings will verify that an improvement has been made.

4.21.5 Test Capacitors

Any well-equipped air conditioning service technician will have at his or her disposal a set of non-polarized motor start and run capacitors to be used as troubleshooting components. Although there are many capacitor testers available for checking these components, it is far better (and easier) to substitute a known good capacitor for a suspected defective one.

When the system fault has been isolated to a compressor or motor that does not run properly, the best way to exonerate the existing capacitor in the unit is to temporarily substitute another one of similar value. This quick and easy method of troubleshooting will save the service technician much valuable time.

At the very least, the test capacitor set should include 4 microfarad, 15 microfarad, and 40 or 50 microfarad units, all rated at 370 volts AC or more. For troubleshooting systems which operate from a 208/240 power line the capacitors should be rated at 440 volts AC or more.

Additionally, it is handy to include a very large value non-polarized capacitor, about 100 to 250 microfarads, in the test kit to provide a method of hard starting a reluctant compressor. The voltage rating of this component should be at least 370 volts AC.

4.22 Air Conditioner Problem Analysis

When the A/C technician is called upon to service a unit, the first step in the diagnostic procedure is to evaluate system performance before any testing or work is done. If a problem is detected, a flow chart can be used as a logical and orderly guide to the general source of the problem. Much information about system operation can be obtained by sight, sound, and touch, all without the use of any instrumentation whatsoever.

4.22 .1 Performance Check

A more formal check of the system can be made by using a thermometer to measure the dry bulb temperature differential of the evaporator inlet and outlet air stream. While such a measurement should take into account the ambient relative humidity, the temperature measurement alone may help determine if a system is not operating properly. Most air conditioning systems will be able to sustain at least a 15 degree F (8.3 degrees C) air temperature difference across the evaporator coil.

Manufacturers of air conditioning units sometimes supply technical performance charts and data which pertain to a particular model. This data may be in the form of a temperature graph which takes into account the evaporator air temperature differential versus the inlet air wet bulb temperature.

Figure 4.18 illustrates a typical performance graph for a window air conditioner. To use the graph, the evaporator inlet and outlet air temperatures, measured with a dry bulb thermometer, are recorded. The evaporator air inlet temperature is measured with a wet bulb thermometer. Using the wet bulb reading, the graph is consulted to determine the specified evaporature inlet and outlet air temperature differential. If the unit's performance is not equal or better to that shown on the graph, the problem must be located and corrected.

As indicated by the typical graph of Figure 4.18, the air temperature differential across the evaporator coil of an air conditioning unit can vary from less than 10 degrees F (5.5 degrees C) to more than 24 degrees F (12 degrees C) in accordance with the wet bulb temperature. The wet bulb thermometer, which can be part of a psychronometer, utilizes the effect of evaporation of water around its sensing bulb to produce a temperature which reading depends upon relative humidity.

Another criterion for air conditioner performance is the power it draws from the AC line. An ammeter can be used to measure line current in amperes, and while this measurement multiplied by the line voltage is not watts, but volt-amperes, it will provide a clue as to the condition of the air conditioner. Window air conditioners generally draw current levels which are very close to the nameplate rating, assuming the unit is operated at a reasonably warm ambient temperature level. A wattmeter, if available, can be used to provide an actual power reading which can be compared to the unit nameplate specifications.

A third method to measure system performance is to connect the manifold gauge set to those air conditioning units which are equipped with access valves. By referring to the manufacturer's specifications of compressor discharge and suction pressure, in accordance with the outside air temperature and humidity, system performance may be evaluated.

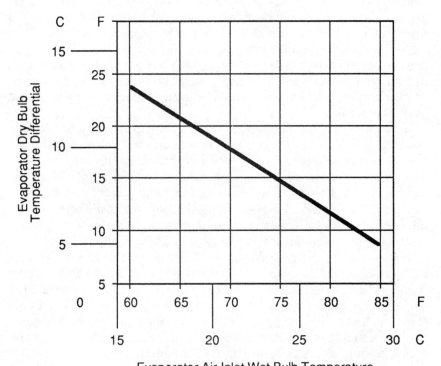

Figure 4.18   Typical Performance Graph

4.22.2 Diagnostic Flowchart

A diagnostic flow chart (Figure 4.19) is one method that can be used to proceed with a logical and orderly method of determining the probable area of fault.   An experienced service technician will automatically have memorized his or her own personal diagnostic technique, similar to that illustrated in Figure 4.19. This will direct attention to the source of the trouble, if one exists. These problems can be broken down into four major categories: mechanical, electrical, refrigerant, or leak problems.

It would not be possible to illustrate here a single diagnostic flow chart which could be used for all air conditioning units. However, due to the ever-increasing complexity of air conditioning systems being produced today and especially those with microprocessor controls,

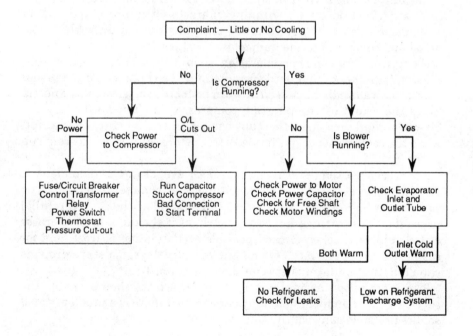

Figure 4.19   Simple Diagnostic Flow Chart

manufacturers are increasingly depending upon flow charts and troubleshooting procedures to lead the service technician in a logical and orderly path to the source of trouble. One might compare this to automobile repair, where service problems of computer-controlled vehicles produced today cannot be easily or properly diagnosed without the use of manufacturer supplied diagnostic information.

4.22.3 Initial Checkout

Certain procedures are necessary when first checking a unit or system for proper performance. All air filters should be inspected and cleaned or changed if necessary. Electrical connections, such as a power cord and receptacle, should be in good condition. The system thermostat, if so equipped, should be set to the lowest setting to ensure compressor operation.

The unit may be turned on and allowed to operate for several minutes (if possible) while certain running characteristics are noted. If an ammeter has been connected to the unit, line current draw should be noted and compared to the nameplate specification.

Compressor and blower operation should be verified. On a split system, both indoor and outdoor units must be checked. Any unusual noises, such as might be caused by worn motor bearings, should be noted. Since the sound of an operating compressor is often masked by the blower, compressor operation can be noted when first starting the unit, when the compressor starts immediately before the blower has come up to speed.

Air discharge temperatures, from both evaporator and condenser, should be checked using either a suitable thermometer or by feel. The compressor high-pressure discharge tube should be checked (carefully) to see if it feels warm or hot to the touch, as it should. The evaporator coil should be checked for any possible accumulation of frost, and for coolness over its entirety. On all but very dry days, moisture droplets should be accumulating between the evaporator fins.

The information assimilated by the above tests should enable the service technician to immediately determine if there is a problem and if so, the probable cause of the trouble.

4.23 Mechanical Problems

4.23.1 Blower Motor

Air conditioner repair problems which are mechanical in nature are usually associated with the evaporator or condenser blower motor assembly. Common faults are blower motors which are sluggish in operation or fail to rotate at all. The usual cause for this condition is dried out bearings, and the cure is to provide the necessary lubrication to restore normal operation. Motors that are equipped with lubricating ports should be oiled annually with a few drops of non-detergent SAE 20 weight oil.

Many blower motors that are used on small window air conditioners do not have any provision for lubrication. They are supposedly "lubricated for life," but the manufacturer of the motor may have a different view of motor life than the person who bought the unit. It is not uncommon for such motor bearings to fail after just a few years of intermittent operation. The situation is further aggravated by the usual dusty conditions under which many air conditioners are operated.

The condition of motor bearings can often be determined by disconnecting power from the unit by means of the line cord or circuit breaker and physically rotating the fan blade by hand. The fan should rotate freely and should free wheel for some time before it comes to a halt. Additionally, there should be some end play in the motor shaft, which can be determined by moving the fan blade back and forth in an axial direction.

If the motor shaft is frozen or seems sluggish, the only cure is lubrication. Sometimes it is possible to correct this condition on lubricated-for-life motors by applying oil on the shaft near the bearings, and working the oil inward by rotating and moving the shaft back and forth. In extreme cases, where there is no end play at all, a judicious hammer blow to the end of the shaft can sometimes restore sufficient play to free up the bearings. A very small quantity of a penetrating oil product, like WD-40, can often get the parts free enough so that lubricating oil can be coaxed into the bearings while rotating the shaft by hand. Sluggish motors will sometimes respond to a higher-than-normal voltage, such as that provided by a Variac, to allow self-operation of the motor as the oil is absorbed into the bearings.

Trying to restore a frozen or sluggish motor back to service might be a time-consuming operation, but it can be far easier (and cheaper) than motor replacement. Another point to remember is that air conditioner repair is almost always done under stress conditions (such as during a heat wave), and restoring the original motor to service can often be done on the spot; replacement may take days or weeks.

If the motor shaft is free to rotate but the motor does not run at all or is sluggish, the problem is electrical. This will be addressed later in this chapter.

4.23.2 Worn Bearings

Motors which have seen many years of service without proper lubrication will exhibit problems which can be attributed to worn bearings. One of the first symptoms to manifest itself is noisy operation, which can be more pronounced at one speed than another. In extreme cases, a motor with bad bearings may have interference between the rotor and stator and cannot develop sufficient torque to start rotation. A manual push of the fan blade can sometimes initiate rotation of the stalled motor.

Worn bearings (or a worn shaft) can be checked by disconnecting power from the unit and working the fan blade back and forth to determine the amount of radial play in the shaft. A motor with bearings in good condition should exhibit essentially no radial play in its shaft. End or axial play is a normal and desirable characteristic; it should, however, be limited to a minimum amount.

It is usually not economically feasible to replace bearings in motors used for window air conditioner service. Replacement of the motor itself is the only practical cure. Larger motors which are part of commercial air conditioning systems can usually be refurbished with new bearings. Fractional horsepower motors which have good bearings but worn shafts can sometimes be restored to service by repositioning the shaft in the rotor assembly to bring a fresh section of the shaft into the bearings.

4.23.3 Fan Blade

Often, a blower motor will not rotate because of interference between one of the fan blades and shroud. The clearance between these two items is often extremely small in some units. This condition can be brought about by mechanical failure of motor bearings or mounts, or by rustout or damage to the sheet metal structure of the unit. A severely unbalanced fan blade can also result in interference problems due to excessive vibration.

Another cause of fan blade interference is mechanical distortion of the cabinet due to improper installation, or damage from abuse of the unit. Correcting this type of problem requires reforming the structure and restoring proper clearance of the moving parts.

Other mechanical problems in larger or commercial air conditioning units may be encountered. These include such items as worn out or slipping belts, which can be adjusted or replaced as necessary. Units which are belt-driven employ bearings in the fan blade assembly, and these items, as well as the motors which drive the blowers, will usually require periodic lubrication.

4.23.4 Compressor

The compressor, being an electromechanical device, can fail either through a mechanical or electrical problem. Either of these conditions will result in a "stuck" compressor. If the problem is electrical, as might be caused by a burned out or open winding, the unit generally cannot be repaired and must be replaced. A compressor that has become stuck through a defect in its moving parts can sometimes be restored to service by a technique which causes the rotor to turn in the opposite direction to clear the obstruction. This method, however, is not successful on compressors which are worn out or have serious internal damage.

4.24 Electrical Problems

Electrical problems in an air conditioning system will be evident when the unit is totally shut down, or runs with one or more inoperative components, such as the compressor or blower motor. Before attempting to troubleshoot any system which appears to have an electrical problem, verify that proper power is being fed to the system. Units which are powered by 208/240 or higher voltage, and those driven by three phase power sources, should be checked for proper voltage from each leg of the power line to neutral or ground. Line-to-line voltages on three phase systems should also be checked. Phase sequence in three phase systems is important; it determines the direction of rotation of all three phase motors and compressors in the system.

4.24.1 Compressor

An air conditioning unit may have an inoperative compressor for any one of many reasons. Should the compressor be stuck mechanically, the locked rotor current that it draws upon start-up will actuate the overload relay, or possibly trip a circuit breaker or blow a fuse. This symptom will lead directly to the compressor for further troubleshooting.

Should this be the case, the best way to determine if the compressor has indeed failed is to try replacement of the compressor start capacitor if so equipped. If the fault is not corrected by substitution, the only way to absolutely prove that the compressor is bad is to disconnect all wires which feed the compressor input terminals (including the start capacitor), and feed AC power directly to the terminals through external wiring, using a new capacitor. The circuit is illustrated in Figure 4.20. The overload relay may be left out of the circuit to eliminate this component as the possible source of trouble. If the compressor does not start and run immediately after power is applied, it has failed.

The service technician should be aware that removal of the protective cover over the terminals of a compressor represents not only an electrical shock hazard but also a risk from an expelled terminal due to pressures from within the system. It is recommended that the protective cover be replaced before applying power to the compressor.

Further investigation of failed compressors is normally not necessary, but if desired the winding resistance and leakage to ground can be checked. A megger is the best way to check leakage. These tests may reveal the internal problem of the compressor. Additionally, analysis of the oil charge of a defective compressor can often help determine the cause of failure.

When the compressor is inoperative and the overload relay or circuit breaker does not trip, the cause is usually some other component or problem and not the compressor itself. If the system employs a relay or contactor, the service technician can often manually activate the relay with a plastic or wooden tool to see if the compressor will run. This will indicate if lack of power to the relay coil is the problem.

If the compressor does not receive power when the thermostat is set to a sufficiently low temperature to call for cooling, the service technician must troubleshoot the thermostat circuit, and low-voltage relay coil circuit, using standard electrical troubleshooting techniques. The low-voltage transformer, in a unit so equipped, should be checked for proper output voltage. Using a voltmeter to trace power through the thermostat circuit may quickly find the fault. If there is a reset push button wired into the system, this should be checked first, as well as all protective components. Loss of all refrigerant charge will prevent compressor operation on those units equipped with a low-pressure cut-off control.

Many air conditioning systems employ one or more pressure controls which are designed to open the relay coil circuit in the event of an abnormal pressure condition, such as too much head pressure or too little suction pressure. Excessive pressure can be caused by an overheated condenser coil or overcharging. Low pressure may be caused by loss of refrigerant or a clogged expansion valve or capillary. A manifold gauge set connected to the service ports of the unit will indicate if there is excessive or insufficient pressure in the system.

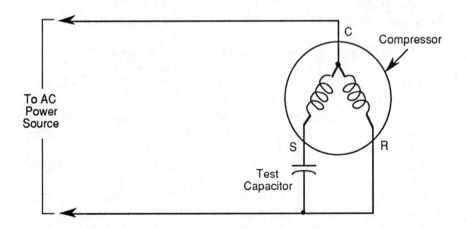

Figure 4.20   Compressor Test Circuit

4.24.2 Induction Motors

Electrical failure of an induction motor will manifest itself either as an open or shorted stator winding, or excessive leakage to ground. An open winding is readily detected by means of an ohmmeter check, but shorted windings (usually partially shorted) are not. It would be prudent to first eliminate the motor capacitor (if so equipped) as the possible source of trouble, by temporarily substituting another one in its place.

Similarly, capacitor start motors which fail to rotate may have inoperative centrifugal switches that must be investigated by disassembly of the unit. Defective centrifugal switch parts are easily replaceable items, and once so replaced will allow the motor to be returned to service. As with the PSC-type induction motor, the start capacitor should also be checked as the source of the problem.

Motors which have shorted windings or leakage to ground will not run at all, may run if manually started, or may run at a slower than normal speed. Additionally, multispeed motors which have one defective stator winding may operate properly on one speed but not on another.

Many times a motor will appear to be good when it is first operated from a cold start. However, as the motor windings heat up, the insulation of the windings will break down, slowing or stopping the motor entirely. For this reason, if a defective motor is suspected, it should be operated for at least an hour to determine if its internal heat buildup will cause a failure.

Often, a motor with a defective winding will operate at what appears to be normal speed, but will overheat. Some motor designs include an automatic reset internal thermal cut-off switch which breaks the electrical circuit at a predetermined temperature. This causes the motor to stop when overheated, and it cannot be started again until it cools down. If the cause of overheating is due to mechanical reasons, such as dry bearings, correction of the problem will allow the motor to be returned to service. Otherwise, overheating due to electrical breakdown of the windings requires replacement of the motor.

Generally speaking, motors with open or shorted windings are not economically repairable. The cost of a replacement motor is less than the labor and materials that would be required for repair. In the case of very large motors which develop shorted windings, it may be possible to consult a professional motor repair house to determine the cost of repairs versus replacement.

4.25 Refrigerant Problems

By and large, refrigerant problems are a major cause of poor air conditioner performance, or total failure of the cooling system. Refrigerants, under high pressure within a system, will eventually leak out to some degree during the lifetime of most air conditioners. A minor loss of refrigerant can be tolerated by some air conditioning systems, but a substantial loss will require service and repair to restore original performance to the unit.

Whether or not to classify an air conditioner as having a "leak" can be a gray area, and subjective in nature. Everyone will agree that an air conditioning system which loses its refrigerant in a month has a leak that must be located and repaired, but how does one classify one which loses its refrigerant in five years?

The criterion for attempting to locate extremely small leaks must be consideration of the cost of periodic recharging versus the time and material required to locate and repair the leak. This is a very real world situation in automotive air conditioning systems, in which some vehicles may require recharging more than once a season while others as little as once every eight or ten years. Compounding the problem is the ever increasing cost of R-12, which is subject to an escalating tax scale. Many leaks which were ignored in the past will have to be corrected.

It is obvious that if the cost of recharging is very much less than the repair costs, the customer will probably live with the problem and have the system recharged as often as necessary. Unfortunately, the dissipation of certain Freon refrigerants into the atmosphere from such air conditioning systems does not benefit the environment. New taxes, placed on CFC refrigerants, may help convince some to repair rather than recharge.

Loss of refrigerant in an air conditioning system will result in less-than-normal performance, which may not be obvious during mild weather. A system with a low refrigerant charge can be detected by several methods.

Larger air conditioning systems, such as central units, often employ a sight glass which permits a visual examination of the flow of liquid refrigerant. A steady stream of gas bubbles usually indicates a low refrigerant charge. One or two bubbles a minute, especially during cooler weather, does not necessarily indicate a low charge.

A pressure check of the system may also be used to determine if the refrigerant charge is sufficient. Air conditioning units which employ capillary tubes as the restricting device should operate with compressor suction pressures that preclude the formation of frost on the evaporator coils. This means, for example, if R-22 is the system refrigerant, the suction pressure should not be much less than 57.5 PSI. This is the boiling pressure of R-22 at 32 degrees F (0 degrees C). Suction pressure can be slightly below that of freezing level since the heat load on the evaporator is sufficient to prevent frost. The temperature scale of the manifold gauge indicates the refrigerant boiling temperature in the evaporator. It is best if suction pressure readings are compared to data provided by the manufacturer of the unit.

A third method to determine insufficient refrigerant charge is to physically examine the evaporator coils after sustained unit operation. With the evaporator blower at full speed and clean air filters in the unit, there should be no buildup of frost. This test must be made at a reasonable outside ambient temperature, such as 75 degrees F (24 degrees C) or more. Some units will normally develop frost when operated at low outside ambient temperatures.

Refrigerant problems are not limited to insufficient charge. Contamination of the refrigerant with air or moisture will prevent proper operation. This can be caused by opening the system to the atmosphere, or service by inexperienced personnel. Contamination problems in some units can also be caused by air or moisture entering the system at any point of leakage where the operating pressure falls below zero PSIG (a vacuum).

Moisture in an air conditioning system can cause breakdown of the refrigerant or oil, and in some cases form ice which restricts refrigerant flow. The cure for such a condition is to thoroughly purge the system with dry nitrogen and add or replace a drier which will trap and hold harmless any residual moisture. In extreme cases, where a system has been left open for a substantial length of time and has accumulated a large quantity of water, several filter/drier replacements may be necessary.

To purge a system, low-pressure nitrogen is applied to an access valve while another valve or connection is opened to allow the gas and moisture to pass through and out of the system. This technique is illustrated in Figure 4.21. Refrigerant R-11 may also be used as a purging agent, but it is strongly recommended that it be reclaimed rather than dissipated into the atmosphere. If the system refrigerant restricting device is not a closed type, such as a capillary tube or expansion orifice, the nitrogen gas or R-11 can be forced through the entire system by applying it to the high-pressure port while opening the low-pressure valve to allow the purging agent to exit the system.

After purging, the vacuum pump is connected to the manifold gauge set and operated until the maximum vacuum developed by the pump is reached, as indicated by the low-pressure gauge. Any system which has an appreciable amount of moisture remaining in it prior to evacuation will require more time to be fully evacuated.

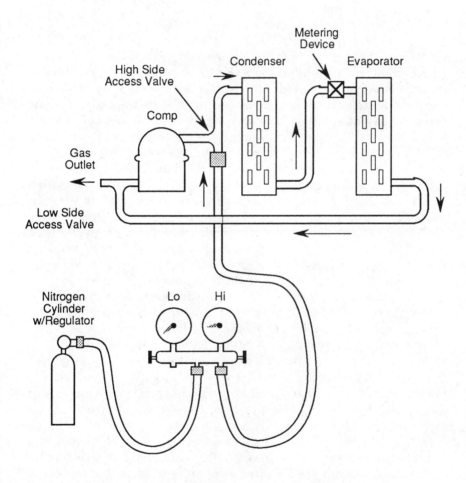

Figure 4.21   Purging A System

If the vacuum pump is not capable of attaining at least 29.5 inches of vacuum (about 100 KPa), the evacuation procedure should be performed more than once, and preferably three times. Each time between evacuations the system is brought up to zero pressure with refrigerant or nitrogen. After the third evacuation the system may be recharged with refrigerant, and its performance checked.

4.26 Leak Problems

All air conditioning systems will eventually lose part or all of their refrigerant charge due to leakage, since internal system pressure is greater than atmospheric pressure. Fortunately, many air conditioning systems are manufactured sufficiently tight so that excessive refrigerant leakage is not a problem during the lifetime of the equipment. Leaks as small as 1/2 ounce per year, while not catastrophic in nature, can be located by sensitive electronic detecting devices.

It cannot be overstressed that a visual inspection of the system can often locate the source of the leak, which manifests itself as an oil stain on what would otherwise be a dry surface. The oil charge within an air conditioning system travels through all components with the refrigerant, and will often leak out at the same time. The refrigerant dissipates into the air, but the oil remains as a telltale sign of a leak.

Any air conditioning system which has zero refrigerant pressure when first encountered will most certainly have a relatively large leak unless it has been opened to the atmosphere. Recharging such a system without repairing such a leak is an exercise in futility; the new recharge will quickly dissipate into the atmosphere.

There are several different methods of locating refrigerant leaks. The method employed often depends upon the magnitude of the leak. It is helpful to know beforehand if the system has a relatively large leak (dissipating its charge in a week or less), or a minute leak which may take a month or a year to affect system operation. This can be determined by learning the history of the system, determining, for example, when the system was last charged.

Additionally, connecting the manifold gauge set to the system will indicate the presence or absence of refrigerant and give a rough idea as to the percentage of charge. By referring to the pressure/temperature scale on the pressure gauge, the presence of liquid refrigerant in the system (usually a normal condition) can be verified. The presence of liquid refrigerant indicates that most of the charge has probably been retained. Alternatively, a very low or zero pressure reading will indicate a relatively large leak.

Before connecting the manifold gauge set to the service valves, check them for refrigerant leakage. It is always possible that the last service technician (or owner) has left the valves improperly sealed which resulted in a refrigerant leak. This simple precaution can save many hours of troubleshooting time.

To perform a leak test, the system needs to be somewhat charged. For large leaks, the pressure within the system can be as small as 5 or 10 PSI, and a soap solution (or commercially prepared leak detection product) will indicate bubbles at the point of the leak. The pressure within the system can be adjusted by controlling the amount of added refrigerant.

Leaks that cannot be located by visual inspection or a soap solution will require more sophisticated methods. This will require a halide or electronic leak detector, either of which is capable of locating extremely small refrigerant leaks.

When a halide torch is used to locate leaks that are relatively large, it would be prudent to charge the system to a pressure only large enough to produce a green flame when the air hose is brought near the leak. Often, all that is required is a pressure as small as 5 or 10 PSI. When large amounts of refrigerant contaminate the atmosphere, the flame will assume a constant green color, making leak detection more difficult.

When performing a leak test using either the halide torch or an electronic detector, every inch of the sealed system must be checked, starting with the most obvious places such as brazed or mechanical connections of one component to another. The sensing end of the detector should be moved slowly, about one inch (2.5 cm) a second, to allow the detector to respond to any possible refrigerant.

Many times it is not possible to determine the source of the suspected leak, either because it is extremely small or in an inaccessible location. If the system does have a partial charge, it can be fully recharged and monitored to determine the time span (weeks or months) for performance to fall off. This will provide important data on the magnitude of the leak.

In stubborn cases where it is known that the system has a leak yet none can be found, it may be necessary to charge the system to an artificially high pressure (not exceeding 150 PSI) using either nitrogen or carbon dioxide. The higher pressure provided by the nitrogen or carbon dioxide will force the mixture of refrigerant and gas out of the leak at a higher-than-normal rate. This will facilitate the leak test.

When the leak is found, its location and size will determine the method of repair. Many leaks in copper or steel components can be repaired by cleaning the area thoroughly and silver brazing over the affected section. Some leaks will require replacement of a section of tubing, connector, or even an entire component. After the repair has been made, the service technician must repeat the leak detection process to be sure that the leak has been corrected.

5

Repair of Compression Cycle Air Conditioners

5.1 General Information

Compression cycle air conditioning systems are the most common type. These units employ one or more hermetic compressors which are used to pump the refrigerant (Freon or similar product) through the condenser and evaporator. The change of state of the refrigerant transfers heat energy from a temperature/humidity controlled area to the outside. Blower motors driving fan blades are used to move the air through the coil assemblies, and various electrical controls complete the unit.

Whether the air conditioning system is a small 5000 BTU window unit or a 10 ton commercial system, the diagnostics and repairs are very similar. The larger systems are usually more complicated than small window units, and may require more sophisticated troubleshooting and repair techniques. This chapter will provide, in detail, the information needed to perform repairs on residential and commercial compression cycle air conditioning systems.

It is assumed that the reader is familiar with the material which has been covered in previous chapters, and is reasonably equipped with the necessary tools and instruments to properly perform the task at hand. The following sections cover the diagnostics and repair of compression cycle air conditioning systems in sufficient detail for almost any situation to which the service technician will be exposed.

Many air conditioning manufacturers supply service notes which may be found inside the unit when it is opened for service. Larger installations may include a service manual which is given to the owner. A schematic diagram of the electrical wiring is often included with even the smallest window unit, sometimes as a decal placed on the interior of the unit. These service notes and schematics provide important information, and should be referred to whenever possible. A schematic diagram of system wiring will help reduce service time on units which exhibit electrical problems.

5.2 Safety Considerations

Air conditioning units, being electromechanical devices, can expose electrical, mechanical, and refrigerant hazards when opened for repair. When servicing these units, care must always be taken to ensure that accidents or personal injuries are avoided.

As with any device which is operated from a power line, the power input to the unit must be disconnected before servicing any electrical part of the system. The best way to disconnect power is to remove the line cord plug (if so equipped) from the power receptacle. Central and large-size units which are permanently wired to the power source cannot be disconnected this way; the power must be disconnected at a cut-off switch designed for this purpose, or at the circuit breaker or fuse box.

Although power to the unit is disconnected, steps must be taken to ensure that it is not accidentally restored by another person. Placing a warning sign near the point of disconnect can go a long way towards preventing an accident. A better method, if possible, is to lock out the cut-off mechanism.

Under some conditions, it is necessary to work on a system that is under power, such as taking voltage or pressure measurements. To prevent accidental dangerous electrical shock, be sure to use only one hand at a time. This will prevent the possibility of receiving an electrical shock with current passing through the body in the area of the heart, which can lead to electrocution.

Because air conditioning systems contain moving parts which represent a safety hazard, always keep loose clothing, hands, and all other parts of the body away from fan blades and belts when working on a unit. As with any electrical shock hazard situation, power to the system should always be disconnected before attempting to make repairs to electrical parts or wiring.

The handling and use of refrigerants and other gases require certain safety considerations. These materials are supplied in steel containers or cylinders, and are under pressure. Some gases, such as carbon dioxide and nitrogen, will be pressurized to 800 PSI and 2000 PSI respectively and, if improperly handled, can represent a dangerous safety hazard.

When handling refrigerant and other gases, always take special care that the cylinders cannot be accidentally dropped. If possible, use chains or other suitable fastening devices to support them and prevent tipping.

Always use proper tools and fittings to make connections to the containers. In the case of high-pressure gases such as carbon dioxide, nitrogen, and others, a suitable pressure regulator must be connected to the cylinder to reduce the output pressure to a safe value. Be careful when operating the valves and controls.

As a general rule, heat should never be applied to a gas cylinder, since this will raise the pressure and could cause an explosion. Some cylinders are equipped with fusible plugs which melt at 150 degrees F (65.6 degrees C) and are designed to prevent abnormal pressure buildup in the event that the cylinder is exposed to heat. High pressure blow-offs are also included in refrigerant containers. These will automatically vent in the event that the container develops excessive pressure.

Heat may be applied safely to a cylinder during a charging operation, using warm water only, to increase container pressure when necessary. A flame should never be used. During the recharging process the liquid refrigerant boils, causing the temperature (and pressure) of the liquid to be substantially reduced. Placing the cylinder in a pail of warm water to raise pressure is permissible if the water temperature is not above 120 degrees F (about 50 degrees C). This technique is often used to facilitate the charging process.

Caution should also be observed when servicing the air conditioning system, which contains liquid and gas refrigerant under pressure. Goggles and protective clothing should be worn. When using the manifold high pressure gauge to measure system operating pressure, liquid refrigerant will accumulate in the high-pressure hose. This will spew out when disconnecting unless sufficient time is allowed for the refrigerant to return to the system as pressures equalize after the unit is turned off.

A closed system should never be opened unless it is certain that all pressure has been relieved. One way to hasten total discharge of a system after bleeding it to zero pressure is to connect a vacuum pump to the system and draw a vacuum of at least 15 inches of mercury (about 50 KPa). Some systems require substantial time to be fully exhausted, and the process should be allowed to continue until all pressure is relieved.

A restricted line or component can cause some pressure to be retained, which can suddenly escape if that part of the system is opened. Even though a system seems to be totally empty of refrigerant, some pressure may still exist in the event of a stuck valve or restricted line. When disconnecting any refrigerant line or connection, always proceed slowly and be aware of any gas escaping as the parts are separated.

Consideration must also be given to proper ventilation of the work area. Although most refrigerants are non-toxic and non-irritating, they should not be allowed to build up in the atmosphere. Use sufficient ventilation to keep the concentration low. Refrigerant concentrations as small as 20% can cause unconsciousness.

Human tolerance to levels of carbon dioxide is even less—as little as 5% can cause drowsiness and 10% may cause suffocation. Since carbon dioxide and Freon refrigerants are odorless, their presence may be unknown.

Other gases, such as propane and acetylene, are highly inflammable, and should never be allowed to escape into the atmosphere. Fortunately, these gases are not odorless, and will automatically signal their presence. Ammonia gas, being an irritant, should always be cleared of any area before entering.

Although Freon refrigerants are nonflammable, they can produce dangerous phosgene gas if burned in an oxygen-fed flame. When checking refrigerant leaks with a halide torch, the presence of a bright blue flame indicates the generation of phosgene.

Another hazard to be aware of is the propensity of ammonia and oil vapor to explode when ignited. Because of the flammable nature of some of the materials which will be encountered in air conditioning repair, a portable fire extinguisher should be kept nearby. Smoking is definitely out when servicing air conditioning systems.

For cleaning purposes, only approved degreasers should be used. Be sure there is adequate ventilation when using these products. Never use gasoline. The vapors produced are explosive, it is damaging to the skin, and can cause dermatitis. If any problem arises with inhalation of gases, or exposure to cleaning fluids, medical attention should be promptly sought.

5.3 Refrigerants

The refrigerant that is used in any air conditioning unit is usually indicated on the system nameplate, which also may depict the quantity in pounds or kilograms. For most residential and commercial air conditioning systems R-22 (monochloro-difluoro-methane) has been the choice of refrigerant. It also finds use in fast food freezing equipment. The boiling point of R-22 is –41.4 degrees F (–1.6 degrees C). This refrigerant is supplied in a green color coded container.

Vehicular air conditioning systems use R-12 (dichloro-difluoro-methane), as do most commercial and residential refrigerators, freezers, dehumidifiers, and water coolers. The boiling point of R-12 is –21.6 degrees F (–29.8 degrees C). R-12 is supplied in white color coded containers. The use of R-12 as a refrigerant will be phased out this decade. An interim product, R-134A, is an environmentally safe refrigerant presently under consideration for use in automotive and refrigerative applications until better substitutes are developed. R-134A cannot be used in equipment designed for R-12, and must not be used as a substitute.

Air conditioners and refrigeration systems which are designed for special purposes such as extremely low temperatures may employ other refrigerants. Additionally, the use of certain Freon refrigerants is subjected to ever-increasing federal taxes, and will gradually be curtailed due to possible damage to the ozone layer in the atmosphere. Substitute refrigerants will become available as these CFCs are phased out. Before using any refrigerant which is not identical to that indicated on the nameplate of any system, consult the refrigerant or air conditioner manufacturer's instructions as to its suitability for substitution.

In order to avoid mistakes, refrigerant cylinders are color coded, as illustrated in Table 5.1. R-12 is supplied in white containers in various sizes from less than 1-pound (0.45 kilogram), up to 145-pound (66 kilogram) cylinders. R-22 is supplied in similar sizes, but is color coded green. All but the largest size are supplied in disposable containers, which may not legally be refilled. Containers smaller than the 125- or

Table 5.1   Color Codes for Refrigerant Containers

Refrigerant	Chemical Identification	Color Code
R-11	Trichloro-monoflouro-methane	Orange
R-12	Dichloro-difluoro-methane	White
R-13	Monochloro-trifluoro-methane	Pale blue
R-22	Monochloro-difluoro-methane	Green
R-502	Azeotropic mixture R22/R115	Orchid
R-717	Ammonia	Silver

145-pound cylinders are designed so that when in the upright position only gaseous refrigerant will be dispensed. When inverted, liquid refrigerant will be supplied. The large size cylinders (over 100 pounds) may have two valves: one for gas and one for liquid. Liquid refrigerant should never be fed into the suction port of a compressor; to do so may cause irreparable damage.

5.4 Charging Equipment

Minimum charging equipment consists of a manifold gauge set and a supply of refrigerant. For a more elaborate, professional set-up, additional equipment which meters the flow of refrigerant and automatically dispenses the desired amount is available. However, this type of sophistication is usually not applied to field service; it finds application in high-volume production, repair, and charging installations where many air conditioning systems are charged each day.

Although many air conditioning systems, especially residential and automotive types, may not be manufactured with a sight glass, an electronic instrument is available as a substitute. This device "listens" to the flow of refrigerant in the liquid line feeding the evaporator and detects when pure liquid is present. This indicates full charge. As with the automatic charging stations described above, this level of sophistication is not needed since there are several methods by which full refrigerant charge can be determined.

Access to the charging valves of many air conditioning systems will require the use of simple tools. To operate shut-off valves which have square stems, it is strongly recommended that the proper wrenches be used. To do otherwise can easily result in damage to the stems, and difficulty in operating them properly. Air conditioners (other than automotive) which have never been recharged and are not equipped with charging valves will require the addition of at least one valve to the system for evacuation or charging.

5.5 Piercing Valves

The common residential room air conditioner which is designed to be installed in a window is manufactured with a completely sealed system that does not contain an access valve. In order to measure system pressure and add refrigerant, a "piercing" valve (Figure 5.1) is connected to the low-pressure side of the system. This is the connection between the evaporator return pipe and suction inlet of the compressor.

Piercing valves are available in several different configurations, depending upon the manufacturer. They are designed so that the two components of the valve are clamped to the tubing at an appropriate location. Once installed, the valve remains as a permanent addition to the system. A sharp pointed screw or needle, part of the valve assembly, pierces the tubing to allow access to the system. Some valves are designed with a combination piercing screw and valve, which is turned with an Allen wrench. Other designs have spring loaded valves which automatically open when the charging hose is attached.

The valves are always supplied with caps that are placed on the threaded hose connection when servicing is completed. These caps prevent dirt from entering the valve. More importantly, they completely seal the valve and prevent refrigerant leakage in the event that the built-in valve sealing mechanism is faulty. Access valves should never be left uncapped.

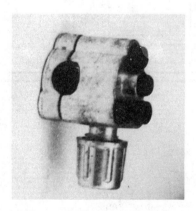

Figure 5.1   Piercing Valve

5.6 Purging and Evacuation

Purging and evacuation are procedures which remove unwanted gases, contaminants, and moisture from an air conditioning system. Whenever a sealed system is opened to the atmosphere for any reason, or has experienced an electrical failure of the compressor (burn-out), purging and/or evacuation is required to restore system integrity.

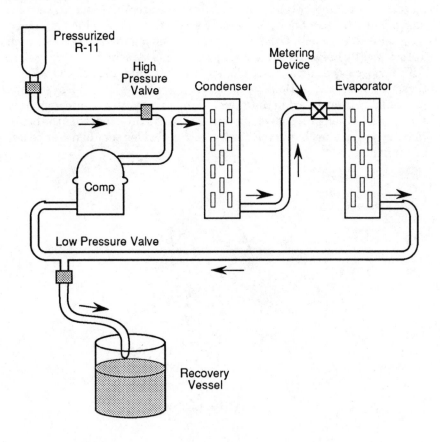

Figure 5.2　Purging a System with Pressurized R-11 Container

Purging consists of applying a neutral gas such as nitrogen or refrigerant, under pressure, to the system at one point while allowing the gas to escape at some other opening. This procedure, illustrated in Figure 5.2, will sweep out unwanted system contaminants. Any system in which a burned-out compressor has been replaced should be purged, since the gases generated by burned insulation or oil should not be allowed to remain in the system. Refrigerant R-11, liquid at room temperature, is often used for this purpose. It will facilitate sweeping out contaminating gases, as well as moisture, in a system which has been exposed to the atmosphere for a considerable length of time. R-11 can be obtained in containers which have been pressurized with nitrogen.

Purging a system may be accomplished by applying the purging gas from a cylinder under pressure to one of the access valves and allowing it to escape from the other. If R-11 is used, liquid is obtained from the pressurized cylinder by holding the container upside down. Using the low and high side access valves as shown in the diagram, this will purge only that portion of the system between the source of purging liquid and gas and the exhaust port, and not the compressor. An alternative method would be to apply pressure to one of the access valves after temporarily disconnecting an appropriate connection or pipe. This will allow the purging gas to flow through any desired portion of the system. After sufficient gas has passed through, the disassembled connection can then be restored.

Using the manifold gauge set for the purging operation allows the cleansing gas or liquid to be easily controlled. It permits the evacuation procedure to be performed immediately afterwards by disconnecting the gas cylinder from the gauge set and connecting the vacuum pump. After the purging operation, an air conditioning system must be completely evacuated to remove the purging agent and all traces of contamination that may remain in the system.

If an air conditioning system has failed "clean," and does not contain contaminants or excessive moisture, it may be evacuated without using a purging agent after repairs have been made. Such a system is partially purged when the service technician applies refrigerant gas under pressure to check the repaired section for leaks, and then performs the evacuation. The vacuum pump will remove moisture and air along with the refrigerant as the system is pumped down.

Figure 5.3 illustrates a setup which can be used to evacuate and charge a system. Small air conditioning systems which do not contain excessive amounts of moisture may be satisfactorily evacuated in one

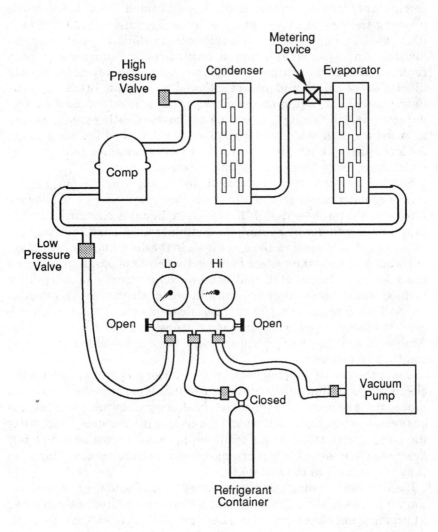

Figure 5.3   Evacuation Set-up Using Manifold Gauge Set as a Connection between Vacuum Pump and System

step if the vacuum pump is capable of reaching a relatively high vacuum, such as 29.5 inches of mercury (about 100 KPa) at sea level. The maximum level of vacuum obtainable is about one inch less for each 1000 feet (305 meters) of altitude. When the vacuum pump is not capable of such a level, or there is considerable moisture that must be removed, a two- or three-step evacuation is necessary. If the air conditioning system has both a low-pressure and high-pressure access valve, the manifold gauge set may be connected to both ports to facilitate the pumping operation.

To perform a multi-step evacuation procedure, the vacuum pump is allowed to operate until the system vacuum is equal to the limit obtainable by the vacuum pump. If moisture is present in the system, additional time will be required for the pump to vaporize the water and remove it. This process will be indicated by the low-pressure manifold gauge reading, which reaches a plateau lower than vacuum pump capability, as the moisture is being removed.

When the vacuum pump limit has been reached, the valve to the vacuum pump is closed. Refrigerant (or nitrogen) from the cylinder is allowed to enter the system until the pressure is zero as read by the low-pressure gauge. The cylinder valve is closed and the vacuum pump valve opened. The vacuum pump will then remove the new charge of gas along with the residual contaminants left over from the previous evacuation.

This procedure may be repeated one more time. The vacuum pump must be allowed to operate until its vacuum limit is attained. The high-pressure manifold valve and vacuum pump valve are closed and the system is then fully charged with refrigerant.

5.7 Recharging

Any air conditioning system which has been opened to the atmosphere, or is suspected of containing contaminants such as air or moisture, should be evacuated prior to recharging and after repairs have been made. Conversely, air conditioning systems may be recharged without evacuation provided it is known that the system does not contain air, moisture, or other contaminants. For example, a window air conditioning unit which has never been recharged, as evidenced by absence of any access valves, may often be brought up to full charge by the addition of refrigerant. An exception to this is a unit which has a leak on the low-pressure side of the system and is relatively low on refrigerant. The suction pressure during compressor operation may have reached vacuum level, drawing in air and moisture through the leak. Such a unit should be repaired and evacuated prior to recharging.

Recharging is a simple procedure, and is always performed through the low pressure side of the system. During the charging operation the compressor is operated so that the refrigerant will flow from its storage container into the relatively low-pressure side of the system. Never attempt to add refrigerant to the high-pressure port when the system is in operation. Such a procedure may cause an explosion of the refrigerant cylinder, and personal injury.

During the charging operation it is possible to monitor both low and high system pressures on those units which are equipped with two access valves. The setup is illustrated in Figure 5.4. All hoses should be purged of air by using either the residual refrigerant pressure in the air conditioning system, or cylinder pressure, to force air out of the hoses. This is accomplished by temporarily loosening the hose connections and allowing a very small amount of refrigerant gas to escape as the air in the hoses is forced out.

It is not necessary to use a high-pressure connection during the recharging procedure as shown in Figure 5.4. If a high-pressure access valve is available, however, it permits measurement of compressor head pressure. This allows monitoring of system operation and can prove beneficial if there is a problem of excessive head pressure. This symptom can be caused by insufficient cooling of the condenser coils.

It is important to note that when the high-pressure gauge is used during charging, the hose will accumulate liquid refrigerant. This represents a safety hazard, since the liquid will spew out when the hose is disconnected. This problem can be avoided through the use of a check valve adapter placed between the high-pressure hose and access valve. An alternate method is to allow the accumulated liquid in the hose to vaporize after the system has been shut down, and before the hoses are disconnected. This is accomplished by opening both low- and high-pressure gauge valves and allowing time for the liquid refrigerant to vaporize.

The charging operation begins with the manifold gauge set and refrigerant cylinder connected to the system as illustrated in Figures 5.3 or 5.4. After all hoses have been properly purged of air, the system pressure is first brought up to cylinder pressure by opening the low-pressure manifold valve and refrigerant cylinder valve. The compressor is then energized by applying power to the unit (with thermostat set to minimum setting), and the suction pressure monitored as refrigerant gas is allowed to enter the system.

Only gas (not liquid) should be allowed to enter the system. Special liquid-to-gas adapters are available to the service technician to allow liquid dispensing from the refrigerant container, but the use

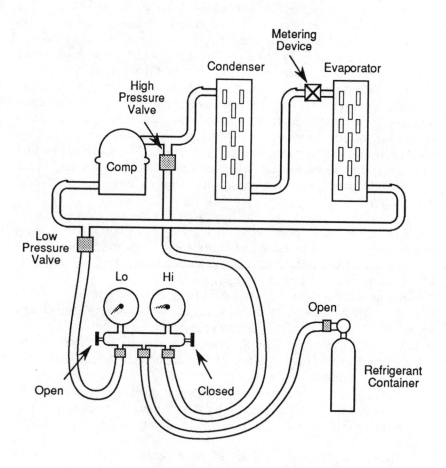

**Figure 5.4 Charging A/C System While Monitoring
Compressor Discharge Pressure**

of these devices is optional. Keep the refrigerant container upright for gaseous refrigerant. During the charging operation the high-pressure manifold valve must always be left in the closed position at all times.

Compressor suction pressure can be checked during the charging process by temporarily closing the low pressure manifold valve. It is important to monitor this pressure when charging units which employ capillary tubes, since the magnitude of the pressure will be a satisfactory indication of the percentage of charge. Air conditioning systems which employ R-22 as a refrigerant will have a compressor suction pressure of approximately 50 to 70 PSI when fully charged.

If the refrigerant container gets very cold during the charging process, its pressure will be relatively low, increasing the time required for recharging. In an extreme case the refrigerant container and air conditioning system pressures will have equalized, and no additional refrigerant will flow into the system. This can be remedied by placing the refrigerant container in a pail of warm, not hot, water to raise its pressure. Never use a flame to heat a cylinder of refrigerant.

Air conditioning systems which employ expansion valves may be equipped with a sight glass which is located in the liquid line feeding the evaporator, usually placed near the compressor unit. Full charge is indicated by the absence of gas bubbles flowing with the refrigerant, indicating pure liquid refrigerant in the line.

When an air conditioning system is fully charged and has been in operation long enough to be stabilized, the temperature of the evaporator return line to the compressor can also be used as an indicator of full charge. The difference in temperature between the return line and the boiling temperature of the refrigerant (indicated by the appropriate scale on the low-pressure gauge), is called superheat, and may be specified by the manufacturer of the system. Fully charged air conditioning systems will usually have evaporator return lines which feel cool to the touch, and may collect condensation on the outside.

Small air conditioning units can be checked for full refrigerant charge by monitoring the line current feeding the unit. At full charge, this current will be very close to the nameplate specification at outside ambient temperatures of about 90 degrees F (32 degrees C). Some manufacturers place service data sheets inside the unit, and these may provide line current information in accordance with the outside air temperature and humidity.

5.8 Fractional HP Induction Motors

Fractional horsepower induction motors are used in air conditioning units to provide the necessary air movement through the evaporator and condenser coils. Some units will employ only one motor, which has a double-ended shaft that drives two discrete fan blades. Other, more complex units are designed with two or more fan motors.

Motor horsepower rating will depend upon the size of the air conditioning unit, and can vary from as small as 1/20 HP to 1/2 HP or more. Most small motors used in window air conditioning units employ sleeve bearings, often with no provision for field lubrication. Commercial air conditioning systems usually employ higher quality motors, which often include oil ports for periodic lubrication (usually once a year).

Small fractional HP motors are usually designed so that the flow of air generated by the fan blade is used to cool the motor. These are called open or "air over" motors. Other designs are completely enclosed, keeping dust and dirt out of the unit. These are referred to as totally enclosed non-ventilated (TENV) motors.

If motor replacement is required, determine the voltage, current, and rpm rating of the defective motor if the replacement motor is to be obtained from any supplier other than the factory authorized parts distributor. Sometimes the equipment nameplate will contain pertinent information concerning the blower motor specifications. This will ensure obtaining an exact electrical equivalent of the defective motor. Replacing one motor with another one of a different rpm or too small an HP rating will prove disastrous. Many motors are equipped with thermal overload protection which will shut down operation in the event of overheating.

Fractional HP motors may be single or multiple speed. Additionally, they are categorized as split phase or permanent split capacitor (PSC) types. Many aftermarket motors used for replacement of lower efficiency split phase units are PSC type. The replacement motor may or may not have exact equivalent color coded leads as compared to the original unit. It is best, if possible, to refer to the schematic diagrams of the air conditioning unit and replacement motor to determine the correct connections to be made.

Many air conditioning systems are supplied with schematic diagrams showing all electrical connections to the various components.

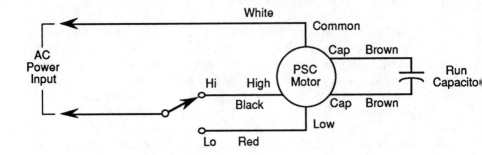

Figure 5.5   Typical Two-speed PSC Motor Circuit

A typical wiring diagram for a two-speed PSC motor is illustrated in Figure 5.5. Manufacturers' diagrams often illustrate the color coding of the wires. This information, if available, will greatly facilitate blower motor replacement in the event that the new motor is color coded differently.

When disconnecting a defective motor, always leave a small portion of the original wires connected to the system. This will provide a reference in the event that a wiring problem is encountered when rewiring the circuit for the new motor.

5.9 Motor Problem Analysis

Motor failure will manifest itself into two categories: electrical and mechanical. The latter will almost always involve dry, frozen, or worn bearings which are generally caused by lack of lubrication. Some motors that are equipped with steel bearings may have worn shafts. Loose bearings or worn shafts will result in excessive radial play, and often cause noisy operation. Small motors with worn components are usually discarded, since the cost of replacement of the bearings will not permit an economically feasible repair.

Motors which have worn shafts can sometimes be returned to service by repositioning the shaft on the rotor. To do this the rotor assembly must be removed from the motor. With the rotor section placed on a support, a soft-head hammer can be used to force the shaft to a new position so that the worn sections of the shaft will not be located at their original positions inside the bearings when the motor is reassembled.

Dry or frozen bearings will result in motors that are sluggish or which do not operate at all. A high-pitched squeal generated by a motor usually indicates that one or more bearings have no lubrication at all. Such motors which require bearing lubrication can often be repaired at far lower cost than replacement.

If the motor is equipped with oil ports, it should be lubricated with a generous amount of non-detergent number 20 SAE oil. Allow time for the oil to seep into the bearing housing while the shaft of the motor is either rotated by hand or coaxed into self-operation. If the motor bearings have not been damaged by the lack of lubrication, the motor can often be returned to service after sufficient oil has penetrated and lubricated the bearings.

If the motor is devoid of oil ports, oil can be applied to the shaft and allowed to seep into the bearings. A far better method is to completely disassemble the motor, remove the rotor from the unit, and soak the bearing housings with oil. The rotor shaft can be inspected for wear and polished with fine steel wool prior to assembly. This method of repair will result in a refurbished motor which will probably operate for as long as it did when new.

A blower motor with an absolutely frozen shaft is not necessarily a candidate for the scrap heap. Working a penetrating oil such as WD-40 into the bearing while rotating the shaft by hand can often result in sufficient play and allow the motor to rotate under its own power. If enough lubricant can be coaxed into the bearings, the motor will be able to rotate under its own power. Of course, the bearings must be fully lubricated with SAE 20 oil before the motor can be returned to service.

If the motor shaft cannot be rotated by hand at all or is sluggish, a judicious hammer blow to the end of the shaft will sometimes break the bearing loose enough to begin the treatment with penetrating oil.

A properly lubricated motor will rotate freely when turned by hand, will start promptly when power is applied, and will coast for some time after power is removed. It is suggested that any motor which has been brought back to life by lubrication be operated for at least an hour to ensure that it has been properly repaired.

One method to check a motor repaired by lubrication of dried bearings is to attempt to start it using low-line voltage as generated by a variable transformer or Variac. This is illustrated in Figure 5.6. A motor in good condition will start (in low speed) under the low torque conditions of 20% low-line voltage. The use of the Variac to temporarily raise the voltage to a sluggish motor is also a valuable technique to get a motor started under its own power.

Blower motors which have freely rotating shafts yet run slowly or not at all will most likely have electrical problems. Windings can be open or shorted. A short circuit between a relatively few turns of the stator coil, or from the stator winding to the frame, can be sufficient to cause excessive motor current draw or prevent a motor from starting or running at proper rpm. Additionally, electrical leakage between turns of deteriorated wire is often temperature dependent. A motor may seem normal when first operated from a cool start, but when it heats up to its normal operating temperature it may slow down or even stop operating. Motors which exhibit electrical faults such as shorted or open stator windings are usually not economically repairable.

Before condemning any motor which is suspected of having a defective winding, check the line voltage being supplied to the motor to be sure that it is at least within   10% of the rated value. Additionally, if the motor employs a start and/or run capacitor, substitution of this component will prove that the capacitor is (or is not) the cause of the problem.

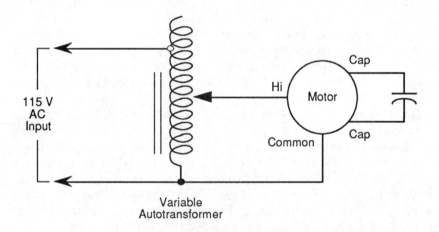

Figure 5.6   Using a Variable Autotransformer to Increase or
Decrease Input Voltage to a Motor

Motor circuits which employ relays or starting switches should also be checked to be sure that these parts are not defective.

Open windings can be easily detected using an ohmmeter to measure winding resistance. Stator coils of single phase motors are connected in series, and the ohmmeter should indicate some resistance from any wire to each of the others. Normal resistance readings of motor windings may be from a few ohms to 100 or more, depending upon the HP rating of the motor. Some PSC motors may have isolated start or capacitor windings which will indicate no connection to the run windings.

An open reading may be caused by a faulty thermal overload switch within the motor assembly, which is usually connected in series with the common circuit of the motor. Any motor which appears to have an open winding may be disassembled to locate the thermal overload (if so equipped). This component may be placed in an accessible location, and replaced if defective.

It generally not possible to determine if a motor winding is shorted by use of the ohmmeter, since short circuits may involve only a few turns of wire and/or be temperature dependent. If the motor does not have an open winding yet fails to run properly, or draws excessive current, it most likely has a shorted winding and must be replaced. Be sure to check leakage to the frame, which should be at least 20 megohms. A megger is the best instrument to use for this purpose.

5.10 Motor Replacement

Blower motor replacement in an air conditioning unit can run from the simple to the very complex, depending upon the size of the system and its design. As a rule, the small window type units can be the most difficult to service, since these are designed to be as compact as possible. Some manufacturers' philosophy is that once a fan motor has failed the unit is not economically repairable, since the cost of a new motor plus the labor to replace it can often approach the discount price of a new unit.

Nevertheless, it is usually possible to remove a blower motor from a unit by performing some amount of disassembly. At the very least, the fan blades must be disconnected from the shafts. These may be secured with Allen set screws, which may be accessible only with special Allen wrenches which are long enough to reach the hub of the fan blade from the outside. Some air conditioning units employ all-plastic fan blades which are secured with clamps instead of screws.

In a worst case scenario, such clamps may be accessible only by removing or pushing to the side one or more fins of a squirrel cage blower wheel, which must be epoxyed back in place during reassembly of the unit.

Most small air conditioner designs permit removal of the blower motor by temporarily repositioning the condenser coil assembly. This is accomplished by removing all screws and/or clamps which hold the condenser in place, and very carefully rotating it, only as much as necessary, out of the way of the motor and fan blade. Note that the two (or more) refrigerant tubes are positioned so that the coil may be moved without kinking or damage to the fragile lines. When performing such an operation, the condenser assembly must be fully supported on the workbench so that it does not accidentally twist and damage the tubing.

Blower motors are fastened to the structure of the unit by various means, and the replacement motor may or may not have the exact same mounting bracket. If the new motor has been obtained from an author-ized parts distributor, it usually will contain an adapter kit (if necessary) to permit proper assembly of the motor into the air conditioner. If the new motor must be obtained from any one of the aftermarket distributors, obtain one which has a compatible mounting bracket for the unit under repair. Many air conditioner units are designed with multiple mounting holes drilled in the motor support assembly, which allow mounting two or more different motor styles.

Be very careful when wiring the replacement motor to the original circuit, especially if it is color coded differently. A miswire can easily result in the burnout of the new motor—a costly mistake. The common lead of the blower motor is usually color coded white, and must be connected to one side of the power line (possibly through the power control switch). Line power must never be applied between the high- and low-speed connections.

If the replacement motor has more speeds than the original, use high speed for single speed units, or high and low for two-speed units. Any wires that are unused should be taped off so that there is no possibility of a short circuit to the chassis or any other component. All wires should be secured to prevent any possible interference with the moving parts of the unit.

After assembly of the motor into the unit and restoration of the condenser coils into the proper position, check carefully for proper clearance between the fan blades and shroud. Such clearance is usually very limited, and a poor assembly job may result in fan blade interference with the sheet metal of the unit.

When operating a unit for the first time after motor replacement, monitor power line current. Any possible excessive current due to a miswire can quickly be turned off to prevent damage to the new motor.

5.11 Thermostat

Thermostats may be checked using an ohmmeter or continuity checker. Over the normal operating range of most air conditioning thermostats about 65 to 85 degrees F (18 to 29 degrees C), the main contacts of the control should open when set to minimum and close when set to maximum, assuming the ambient room temperature is somewhere within the range of the thermostat.

Some thermostats have additional contacts which are used to operate the fan blower motor. These contacts may operate simultaneously with the main contacts, or be offset by 1 or 2 degrees. Checking continuity of the switching contacts as the control is rotated over its range will usually indicate whether or not the part is functioning properly.

Another way to check a thermostat is to subject it to temperatures above and below its setting, and checking the contacts for closure and opening. If the thermostat is equipped with a capillary tube for temperature sensing, it is a simple matter to cool it using cold water or ice, and warm it using heat from the hand.

Thermostats are factory set to a range and differential which the manufacturer deems correct, but this setting may or may not be suitable for the application. For example, many thermostats may close only at temperatures above 70 degrees F (21 degrees C), and this prevents the unit from being operated at relatively cool ambient temperatures when high humidity levels require air conditioner operation.

The range of the thermostat, and often its differential, are sometimes field adjustable by turning small screws inside the mechanism. A cover may first have to be removed to gain access to the adjustment screws, or sealing compound removed. Only a small adjustment is usually necessary. Differential settings should generally not require readjustment, since attempting to set too narrow a differential may cause short cycling of the compressor.

There are many styles of replacement thermostats. Any part which will fit into the unit, has the correct sensing mechanism (bimetallic or capillary tube), and is rated for the same or greater current and voltage range of the original may be substituted.

5.12 Hermetic Compressors

The compressor is usually the most expensive component in an air conditioning system. When the compressor does not operate, the fault may or may not lie with the compressor, and only a thorough checkout will prove if replacement is necessary.

When an air conditioning unit is powered with the thermostat adjusted to its minimum temperature setting, the compressor should operate providing the ambient temperature is above the thermostat setting by at least a few degrees. A compressor which is mechanically stuck or has a shorted winding will usually cause its overload to cut out. Sometimes the circuit breaker or fuse protecting the circuit will trip or blow. Monitoring the line current drawn by the air conditioning unit can provide information which will aid in diagnosing the fault.

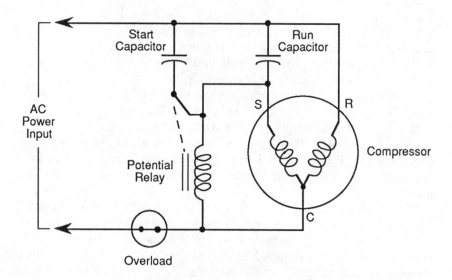

Figure 5.7   Typical Compressor Wiring Diagram Showing
Optional Hard Start Circuit

Failed compressors with open windings are not as common as those which have shorts or mechanical problems. When this is the case, the line current to the system will remain substantially below normal when the unit is turned on.

Figure 5.7 illustrates a typical compressor circuit which includes a potential relay and start capacitor. Many air conditioning units delete the relay and start capacitor from the circuit. Compressors used in either type of circuit should first be checked by measuring the line voltage applied between the common and run terminals of the unit to be sure that it is within 10% of the rated value. Additionally, the run capacitor, connected to the start winding, can be replaced with a new unit to be certain that this component is not the cause of the fault. A resistance reading can be taken between the terminals of the compressor, and between any winding and case, to be sure that there is no open, short, or leakage to ground. This test must be made with all wires removed from the compressor terminals.

If all of the tests performed point to a failed compressor, there is only one way to be absolutely certain that it has indeed failed. This is to disconnect all wires from the compressor terminals and wire the compressor to the power line using a new capacitor which is known to be in perfect condition. If the compressor fails to run with this test circuit, it is bad. The protective cover should always be placed over the terminals, before applying line power, to prevent injury from a possible expelled terminal.

A megger can be used to predict remaining compressor life. This method involves the use of an instrument called a megger, which is capable of determining electrical leakage between the windings and shell, as illustrated in Figure 5.8.

The megger impresses a high voltage, usually 500 volts, between the compressor stator winding and frame to determine the amount of current flow and electrical leakage, measured in megohms (millions of ohms). An ordinary ohmmeter, which employs a very low voltage for resistance measurements, cannot properly determine leakage in a hermetic compressor. One must be careful when using a megger, since its 500-volt output represents a shock hazard. Additionally, hermetic compressors must never be tested with a megger when the air conditioning system is under vacuum conditions.

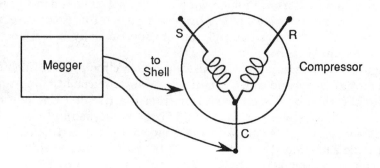

**Figure 5.8   Using Megger to Check Electrical
Condition of a Compressor**

It may seem that the resistance between the windings and frame of
a compressor would be infinite, but in the real world of oil- and
refrigerant-drenched windings, this resistance is a measurable,
finite value. Table 5.2 illustrates the probable condition of a typical
hermetic compressor, as determined by the leakage resistance of its
windings to frame.

It is best if the compressor is checked when fully warmed up after
operation for 10 minutes or more. This will help drive out refrigerant from the windings, which might cause a lower than normal
reading. All wires must be disconnected from the compressor
terminals prior to the megger test, to prevent a false reading from
some other part of the system. The connector block of the compressor
should be thoroughly cleaned before taking a measurement, to prevent a false reading due to dirt and grime.

Readings of 50 megohms or less indicate a problem in the system.
This may be caused by moisture, and such systems should be fully
evacuated after replacing or installing a drier. If this procedure does
not improve the resistance reading, the oil in the system may be
contaminated. Commercial air conditioning systems may have pro-
visions to change compressor oil if contaminated.

Readings of less than 20 megohms taken after thorough evacuation
and recharging indicate severe system conditions, and compressor life
may be relatively short.

Table 5.2   Compressor Condition as a Function
of Megger Resistance Reading

Measured Resistance	Probable Condition
100 Megohms or more	Excellent; no preventive maintenance required.
50 to 100 Megohms	Some moisture present; replace drier and evacuate
20 to 50 Megohms	Presence of contaminated oil and/or severe moisture content. Several drier changes necessary.
0 to 20 Megohms	Very severe contamination or deterioration. Change oil and add oversize drier.

If a compressor is in a "stuck" or locked rotor condition due to a minor mechanical problem, it is sometimes possible to correct the fault by electrical means. It is assumed that the service technician has first verified that the cause of the problem is not head pressure due to trapped refrigerant in the high-pressure side of the system.

One method to free a stuck compressor is to temporarily replace the start capacitor with one of much greater value to provide additional starting torque. Another method is to apply a higher than normal voltage to the compressor (such as 240 volts for a 120 volt unit) for a very short interval (1 or 2 seconds) to see if that will break loose the obstruction.

A third method is to run the compressor in a reverse direction by interchanging the start and run connections in the circuit, as illustrated in Figure 5.9. This method may allow the rotor to turn a sufficient amount and push away whatever contaminant is preventing its normal forward rotation.

If the above methods fail to restore the compressor back to normal operation, the unit is beyond repair and must be replaced.

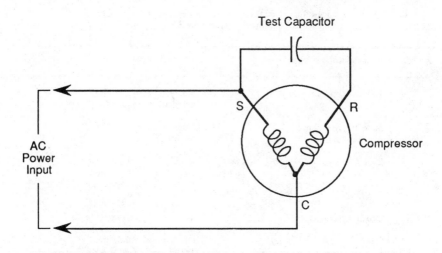

Figure 5.9   Circuit to Reverse Operation of Stuck Compressor

5.13 Compressor Replacement

When compressor replacement is necessary, it is easiest if an exact replacement unit is used, since the inlet, discharge, and electrical connections will be located in the same position as the original. However, due to the multitude of compressors available on the market, plus the fact that many compressors have been discontinued by various manufacturers, it is not always possible to obtain an exact replacement.

If the repair parts supply house is not able to cross reference an equivalent unit, it is possible to select a suitable replacement compressor by matching the voltage (single or three phase), frequency, current rating, and BTU capacity of the original part as illustrated in Table 5.3. Back pressure ratings and starting torque requirements must also be met. The electrical information can be obtained from the compressor nameplate and BTU rating from the unit nameplate. Since the new compressor may very well have a higher EER rating than the original, it is entirely possible that it will draw less current from the line than the original one. The service technician is cautioned to use only compressors designed for heat pump service on such systems. Ordinary compressors are not designed for the extraordinary pumping ratios, and can quickly fail.

It may be permissible to replace a reciprocating compressor with a rotary type, providing that the above restrictions apply. The best source of such information is the factory authorized parts supplier of the unit under repair. When using a different style compressor, be certain that the inlet and outlet connections, and the base, are compatible with the air conditioner under repair. Additionally, the size and shape of the new unit must allow it to fit into the available space.

When replacing a compressor, it is mandatory that the overload relay (if so equipped) and capacitor be matched to the new component. Do not use the original parts if the new compressor is a different model number than the original. Table 5.3 is a typical list of replacement compressors used in air conditioning service.

Before attempting to remove the old compressor, the system must be totally discharged of all refrigerant. Use of a refrigerant recovery system is recommended. Be sure to disconnect the unit from the power source before proceeding, and use goggles for protection. Large commercial-type air conditioning systems sometimes employ shutoff valves which allow the service technician to isolate the compressor from the rest of the system, preserving most of the refrigerant. Install the manifold gauge set to monitor system pressure as the refrigerant is discharged to zero pressure. Allow sufficient time for all   liquid refrigerant in the system to vaporize

Table 5.3   Typical List of Replacement Hermetic Compressors for Air Conditioning Service (Courtesy of Tecumseh Products Co.)

Model Number	Rated BTU Capacity	Power Requirements
AE8458E	5800	115V/5.2A
AE8467E	6700	115V/6.0A
RK5480E	8100	115V/6.6A
RK5510E	10300	115V/8.2A
RK5515E	15500	230V/6.4A
AB5520G	20500	230V/10.8A

and be discharged, which will be indicated by a zero pressure gauge reading when the manifold gauge valves are temporarily set to the closed position.

If the system under repair is not equipped with access valves, one must be added. Use a piercing valve if necessary, installed in the connection between the evaporator and suction inlet of the compressor. The valve should be located where the heat of the brazing operation during compressor replacement will not destroy heat-sensitive gaskets or O-rings in the valve.

The manifold gauge set valves must be left in the open position during disconnection of the tubing from the compressor to avoid pressure buildup. Use an acetylene or Mapp gas torch to heat the compressor connections so that the tubes may be pulled out. After removal of the tubing, allow them to cool and cap immediately to prevent dirt and air from entering the system.

The open connections of the old compressor may be pinched shut and brazed to avoid spillage of any oil that may remain in the compressor. If the failed compressor has burned out, the oil will probably have an acidic, odorous quality. It is recommended that such systems be purged of contaminants after removal, and prior to compressor replacement, by using either R-11 or the same refrigerant that is used in the unit. R-11 can be passed through the system as a liquid, and if it discharges in a clean state the system is properly purged. If the oil charge in the failed compressor is clear and clean, the unit will probably not have to be purged.

Before installation of the replacement compressor, check the location of all connections and be certain that all necessary fittings and tubing are available. New replacement compressors are usually supplied with the proper oil charge. If in doubt, check to be sure. Any compressor operated without oil will quickly fail.

Central or commercial air conditioning systems should be equipped with a new filter drier when the compressor is replaced. The best method is to install two—one in the compressor suction line and another in the liquid line between the condenser and evaporator. This will ensure that any residual moisture present in the system will be absorbed by the desiccant, and any possible solid contaminants will be trapped.

Before performing the brazing operation during installation of the replacement compressor, the ends of the tubing must be thoroughly clean. If they are not, the finished joint will leak and must be redone properly. The manifold gauge set should be connected to the system and the valves left open during brazing to allow pressure buildup from the heat to dissipate. Allowing an inert gas such as nitrogen to flow through the system during brazing will prevent the formation of oxides on the interior of the refrigerant lines. Avoid heating the tubing close to the compressor case. To do so may melt the factory brazed connection.

When the brazing operation is completed, allow the tubing to cool naturally, or use a water-saturated cloth to hasten the process. Use water to clean any residual flux from the joints. This will not only prevent corrosion, but will remove any flux which may be masking a leak.

When the compressor replacement operation is completed, the joints must be leak tested before evacuating and charging the unit. This is accomplished by pressurizing the system with the proper refrigerant and using a halide torch or electronic leak detector to search for leaks. It is best if the system is pressurized to only about 10 PSI at first so that any large leak will be easily detected. If none is found, the system pressure can be increased to 50 PSI or more and very carefully checked at all locations where the parts were reassembled.

If no leaks are found, the refrigerant in the system is allowed to dissipate and the normal evacuation and recharging procedure is followed. When evacuating the system, be sure that it will hold vacuum with all valves closed. This will provide further assurance that the system is leak tight.

5.14 Oil Charge

Hermetic compressors rely on a charge of oil to provide the necessary lubrication that prevents wear and keeps the moving parts working smoothly, much like an automobile engine. Some of the oil mixes with the refrigerant and travels with it throughout the system, always returning to the compressor sump or crankcase. Most systems never need an oil charge unless there has been significant leakage due to a refrigerant leak or broken line. Oil should never be added to a system unless necessary. An excess oil charge will reduce cooling capacity and may damage the compressor.

Some commercial systems have a provision to determine the oil level in a compressor by viewing the level through a sight glass. Others may have a removable plug which, when loosened under refrigerant pressure, will spew out oil, indicating the presence of some oil in the system.

If it is determined that oil must be added to a compressor, use only refrigerant oil of the correct viscosity. This information may be obtained from the manufacturer of the compressor or air conditioning system.

Oil may be added to a system which has only one access valve and is under refrigerant pressure by using a pressurized container of oil, as illustrated in Figure 5.10. However, if the system is not under pressure, a vacuum pump may be used to add an oil charge.

The method shown in Figure 5.10 depends upon the use of refrigerant pressure in the oil container to force the oil into the system. The compressor is operated so that the pressure in the suction line is lower than the pressure in the container. Before starting the procedure, the container of pressurized oil is vigorously shaken to mix oil and refrigerant. The oil container is held upside down to allow the oil to enter the system. Care must be taken to prevent liquid from reaching the suction inlet of the compressor.

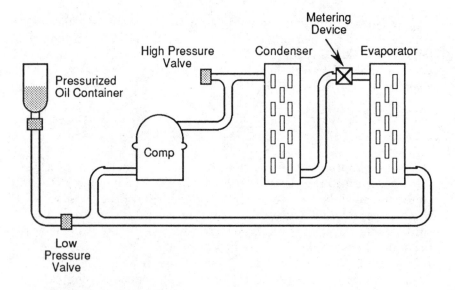

Figure 5.10   Adding Oil to an Air Conditioning System
Using One Access Valve

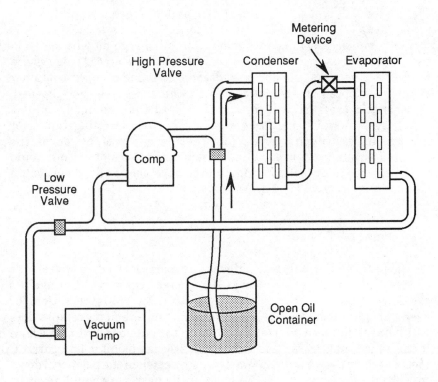

**Figure 5.11   Adding Oil to an Air Conditioning System
Using Two Access Valves**

An alternative method to add oil to a system is to use the negative pressure developed by a vacuum pump. This method can be used on systems which have two access valves and are devoid of refrigerant, and is illustrated in Figure 5.11. It permits the use of an open container of oil. The vacuum created by the pump allows atmospheric pressure to force the oil into the high-pressure side of the system. The oil will return to the compressor when the system is charged and operated.

The open end of the charging hose, connected to the high-pressure access valve, is placed in an open container of oil. When the vacuum pump is operated, the suction generated by the pump will cause the oil to be drawn into the system. This method will cause the air in the hose to be drawn into the system, and when the operation is completed the system must be fully evacuated and charged using standard procedures.

If only one access valve is available, an oil charge may be added in two steps. The system is first evacuated. With the access valve closed to preserve vacuum, the pump is disconnected and an open container of oil substituted in its place. When the access valve is opened, atmospheric pressure will force the oil into the system.

Any system which has been repaired by disconnecting tubing or fittings can be easily charged with oil, if necessary, simply by pouring the required amount into an open line. When using this method, oil should not be applied to the suction side of the compressor. To do so may cause damage as the liquid oil is drawn in.

5.15 Overload Device

Overload relays or thermal cutouts are part of every compressor's electrical circuit. They are designed to protect the compressor against excessive current and overheating due to locked rotor conditions or overload. Most window-type air conditioning units employ compressors which have the overload relay mounted on the outside of the unit where it can sense compressor temperature as well as monitor line current. Most overload devices will automatically reset when the parts cool down to normal temperature. The function of the overload should never be defeated, since operation of a compressor under overload conditions, even for a few seconds, can destroy it.

Larger compressors are designed with the overload relay built into the unit, and these components are not field replaceable. Whenever an internal overload relay opens the electrical circuit to the compressor windings, it is necessary to allow the assembly to cool down before its contacts will close. This may take 20 minutes or more, depending upon the temperature of the compressor assembly.

Field replaceable overload relays are mounted against the compressor shell so that they may properly monitor compressor temperature. If replacement is ever required, the new part should be an exact equivalent of the original. If the wrong part is substituted, compressor protection may be lost. The cover plate should always be installed to ensure proper compressor protection and prevent injury from a possible expelled terminal.

An overload relay may be checked with an ohmmeter or continuity checker. The contacts should be closed when the relay is at normal ambient temperature. However, overload relays with pitted or corroded contacts may present an intermittent condition, and cause the compressor to stall during operation. If an air conditioning unit seems to operate normally most of the time, but sometimes trips the power line breaker, the problem may be an intermittent overload relay. If inspection of the part reveals corrosion or contacts in poor condition, a new relay may be installed to see if that corrects the problem.

5.16 Capacitor

Capacitors are electrical components which draw a "reactive current" from the power line. They are used to provide a phase shift of the current so that the start winding of a single phase induction motor may be excited by an out of phase current. This produces an offset magnetic field which provides the starting (and running) torque for compressors, as well as blower motors.

Single phase compressors always employ running capacitors which are connected between the "run" and "start" windings of the motor. Many blower motors also are designed as permanent split capacitor (PSC) types which are more efficient than the cheaper, split phase designs. These capacitors are non-polarized, and are usually rated at 4 microfarads capacitance or more. In general, the larger fan motors and compressors will have larger values of run capacitors. Many window air conditioning units have two section capacitors, one for each part of the circuit.

Some air conditioning manufacturers also include a "hard start" capacitor and relay in the compressor circuit. Alternate methods include connecting a positive temperature coefficient (PTC) resistor across the run capacitor, as illustrated in Figure 5.12. This component exhibits a low resistance value when cool, and quickly increases in resistance as it heats up during compressor operation. Hard start circuits are used to obtain additional starting torque from the motor by temporarily paralleling the permanent run capacitor with another circuit to increase starting winding current. This provides increased torque which helps the compressor to start during any possible existing head pressure.

When a compressor or PSC fan motor fails to operate properly in a system, first eliminate the capacitor as the possible fault by substituting a new part. If the problem is cleared, the capacitor that was replaced is defective. This method of troubleshooting will often be more expedient than trying to check a capacitor by traditional means.

Capacitors can fail in several ways. They may become open, shorted, or have excessive electrical leakage. Some capacitors have built-in fuses which are designed to blow in the event of an excessive current surge through the device, as will happen if the capacitor shorts out. Such a capacitor will exhibit an open circuit when tested.

Some large value capacitors have an external resistor connected across the terminals. This is a safety feature that automatically discharges the capacitor within a reasonable time after power to the

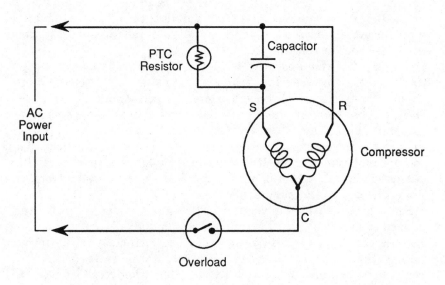

**Figure 5.12   Compressor Wiring Diagram showing
PTC Hard Start Circuit**

circuit is removed. All capacitors without bleed resistors should be treated as though they are charged, since they can easily hold a charge long after the air conditioning unit has been turned off. This represents a shock hazard.

A capacitance meter can be used to measure the value of capacitance, and its leakage, of any unit that is suspected of being defective. However, an ordinary ohmmeter, set to a mid-range scale, can be used to determine if a capacitor is shorted or open.

To make this test, the capacitor should first be discharged so that it has no residual charge. A low-value power resistor, such as 100 ohms five watts, is connected across the terminals for a few seconds. The resistor (and bleed resistor if so equipped) is disconnected and the leads of the ohmmeter connected in its place across the capacitor terminals. A normal indication is for the ohmmeter reading to quickly go to a low resistance reading and then slowly increase to infinity as the capacitor charges up. Capacitors which are shorted or leaky will indicate a finite resistance value that does not change, and those which are open will not produce any reaction at all when first connected to the ohmmeter.

The ohmmeter test described above does not check the capacitor at its normal operating voltage, which may be 120 volts AC or more. By using an ordinary AC ammeter capable of measuring 10 amperes, it is an easy matter to check a capacitor under conditions which more closely resemble actual operation. This method, illustrated in Figure 5.13, also provides a reasonable accurate measure of the value of capacitance.

In this circuit the capacitor is subjected to a potential of about 120 volts AC from the power line, and the current draw is measured by the ammeter. A 10 ampere fuse is placed in series with the ammeter to protect it in the event that the capacitor is shorted. A good capacitor will draw about 0.045 amperes from a 120 volt 60 Hertz power line for each microfarad of capacitance. At 50 Hertz, the current will be 0.038 amperes.

The actual value of the capacitor may be checked against its indicated value by using the expression:

UFD = (Reading in amperes)/0.045 for 120 volt 60 Hertz, or

UFD = (Reading in amperes)/0.038 for 120 volt 50 Hertz

where UFD is the value of capacitance in microfarads.

Any capacitor which does not draw the expected current, within plus or minus 20%, is suspect and should be replaced.

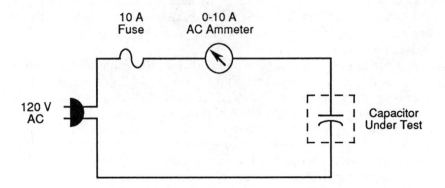

Figure 5.13   Testing a Capacitor Under 120 Volt AC Excitation

5.17 Hard Start Relay

Hard start circuits are used in some air conditioning systems to help avoid the problem of a locked rotor condition when the compressor is called upon to restart without total pressure equalization. This may occur if the thermostat is manipulated, or if there is a momentary loss of line power. Hard start circuits are usually limited to single phase systems, since three phase compressors have far greater starting torque than those which run on one phase. Many manufacturers now employ solid state time delay controls, in both small and large systems, which automatically prevent compressor operation until a predetermined time elapses after the compressor shuts down.

Additional starting torque may be provided by a large value capacitor which is temporarily connected, during the start sequence, across the run capacitor. Once the compressor has reached speed, the extra capacitor is switched out of the circuit, usually by means of a set of relay contacts. A potential, current, or thermally activated relay can be used for this purpose. Other type circuits, such as positive temperature coefficient (PTC) resistors, may also be used in place of capacitors. These circuits operate without the need of relay contact or moving parts, and can be more reliable.

There are two categories of electromagnetic relays: current operated and potential operated. Current relays sense the starting current drawn by the compressor, which may be 10 or more times its running current, and which open a set of contacts when the start sequence is completed.   The coil resistance of a current relay is extremely low, being much less than 0.1 ohm.

A typical current relay circuit is illustrated in Figure 5.14.   When the circuit is idle, the contacts of the relay are open and the start capacitor is not connected to the circuit.   When power is applied between the run

and common leads of the compressor the high inrush current demanded by the compressor, fed through the coil of the relay, causes its contacts to close.   This connects the start capacitor to the circuit.

When the compressor reaches operating speed, its current is reduced, causing the relay contacts to open and disconnect the start capacitor from the circuit.

Potential starting relays differ from current relays in that the contacts are closed prior to applying power to the compressor.   This has an inherent advantage over the current relay, since no arcing can take place at the instant of application of power, as may occur with the normally open contacts of a current relay.   Figure 5.15 illustrates a typical motor starting circuit using a potential relay.

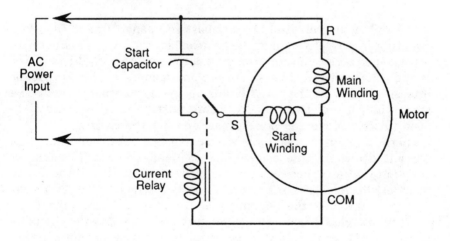

Figure 5.14   Typical Motor Starting Circuit Using a Current Operated Relay

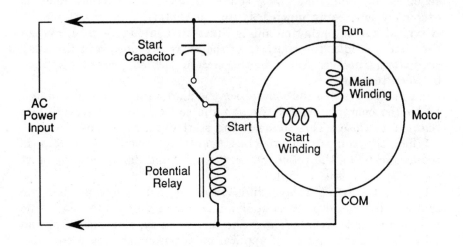

**Figure 5.15   Typical Motor starting Circuit Using
a Potential Operated Relay**

The coil of the potential relay consists of many turns of fine wire resulting in a relatively high resistance. It responds to voltage and is connected across the start winding of the compressor. When power is applied, the closed contacts of the relay connect the start capacitor into the circuit. As the motor builds up speed, the voltage across the start winding increases rapidly. This causes the relay contacts to open, disconnecting the start capacitor from the circuit.

Thermal relays do not have electromagnetic coils, but utilize the current drawn by the compressor to heat a bimetallic strip or resistance wire. Since compressor current must flow through the "coil" of the relay, its resistance must be extremely low as in the current relay. At the beginning of the start sequence the relay contacts are closed, connecting the start capacitor into the compressor circuit. When the relay heats up due to a minute amount of power being dissipated in the relay coil, the contacts open. The thermal relay connection to the compressor and start capacitor is very similar to that of the current relay.

Solid state controls are becoming more and more popular as these parts become lower in cost and more reliable. They may employ thermistors, which are variable resistors that change in value with temperature. Other, more sophisticated controls may employ thyristors (triacs or SCRs) or other solid state components to control the starting circuit of the compressor. More elaborate air conditioners use microprocessor controls which have logic circuits that automatically take into account many operating factors. These devices are miniature computers that control the entire system.

The advantage of solid state controls is the elimination of contacts and moving parts. This results in no arcing or wear-out mechanism as found in electromechanical components. However, they are subject to catastrophic failure, especially when subjected to excessive current or voltage due to a line voltage transient or failure of some component in the system. The service technician will probably replace any solid state module that does not perform properly; field repair is generally not feasible.

Current, potential, and thermal relays may be checked by examination of the contacts for pitting, and the mechanical parts for excessive wear. Contact resistance may be checked with an ohmmeter which is set to the lowest scale. Normal readings should be 10 milliohms (0.01 ohms) or less. The coil resistance of the relay may also be measured with the ohmmeter. Normal readings will be virtually zero ohms for current and thermal relays, and a finite value (several hundred ohms or more) for potential relay coils.

The best way to check relays, as with any electrical component, is under real-world conditions where the component is subjected to the voltage and current it experiences in the system. This may be easily accomplished by connecting an ammeter in series with the starting capacitor so that its current during the start sequence may be monitored. For this application, an analog clamp-on ammeter is best, but other types may also be used. A clamp-on ammeter, illustrated in Figure 5.16, allows this test to be made without breaking the circuit as required when using a conventional instrument.

The ammeter should be set to a range suitable for the expected starting current of the circuit. This will be probably be less than 10 amperes. When power is applied to the air conditioning unit, the ammeter should immediately indicate the surge of motor starting current, then fall to zero as the compressor starts. This sequence will occur very fast, but it will provide verification that the hard start relay or solid state circuit is doing its job.

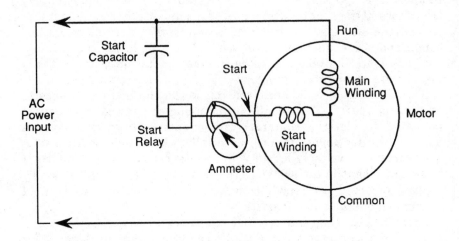

Figure 5.16   Using an Ammeter to Check Starting Circuit

5.18 Pressure Cutoff

Many air conditioning systems other than window units employ high-and/or low-pressure cutoff controls which protect the compressor in the event of a malfunction in the system. These controls are normally dormant during the life of the equipment, and rarely activated. However, they can fail in either an open or intermittent condition and can prevent proper compressor operation.

The high-pressure cutoff control is placed in the high-pressure gas discharge line of the compressor where it can monitor head pressure. The contacts of the control, which are normally closed, are connected in series with the power relay or contactor coil that controls the compressor.

If system head pressure exceeds a predetermined level such as might be caused by failure of the condenser fan motor or blockage of condenser air flow, the resulting excessive pressure will trip the control, opening its contacts and shutting down the compressor. This control may not automatically reset itself when pressure returns to normal; a reset button must be pressed to restore operation.

When a high-pressure control is activated, the cause should be investigated by connecting the manifold gauge set to the system and monitoring the pressure readings. Any obvious fault such as insufficient air flow through the condenser should be corrected. This may be as simple as thoroughly cleaning all dirt and debris out of the condenser coils, or tightening a slipping belt.

The low-pressure cutoff control is placed in the compressor suction line to prevent operation of the compressor in the event of an abnormal condition, such as might be caused by insufficient refrigerant charge. As with the high-pressure cutoff, the contacts of the low-pressure switch are normally closed and connected in series with the coil of the power relay or contactor.

A system which has lost all or part of its refrigerant may not operate at all, or will operate for a few seconds until the low-pressure cutoff interrupts power to the relay coil. A manifold gauge set connected to the unit will indicate suction line pressure levels.

Pressure cutoff controls can fail catastrophically or intermittently, with open contacts. When this happens, the compressor will not operate at all, or will not remain in operation, even though the unit is fully charged and all pressures are normal. If this symptom is encountered, it is a simple matter to isolate the fault to one or the other pressure controls by temporarily connecting a jumper wire across the connections of the suspected control. If the control is defective, compressor operation will be restored.

Pressure controls are usually not field repairable, and must be replaced. Unfortunately, this may require total discharge of all system refrigerant. If available, a refrigerant recovery system may be used to save the refrigerant and prevent it from being dissipated into the atmosphere.

5.19 Capillary Tube Service

The capillary tube is a passive device with no moving parts, and will rarely require service. However, due to its relatively small inside diameter, it is subject to blockage from foreign particles, ice formation, or mechanical damage. A blocked capillary tube will prevent flow of refrigerant throughout the system, which is easily detectable in a fully charged system by checking the condenser temperature. It will feel abnormally cool to the touch, since the compressor suction inlet is starved of refrigerant. A manifold gauge connected to the low side of the system will indicate an abnormally low pressure.

A blocked capillary tube can be checked by operating the compressor for about a minute and then shutting the system down. The normal sound of liquid refrigerant flowing through the capillary tube should be heard. If no sound is evident and the capillary tube is cool to the touch, it may be blocked by ice formation. If so, thawing it out by warming with the hand will restore refrigerant flow, which can be detected by sound.

If blockage of the capillary tube is caused by ice, the cause of moisture contamination should be determined. A small window air conditioning unit may be repaired by a thorough evacuation of the system to remove the moisture. The system should be pumped down to the final vacuum level attainable by the pump, and must be in excess of 29 inches of mercury (100 KPa) at sea level. Allow about 1 inch of mercury less for each 1000 feet (305 meters) of altitude. Evacuation may take 1/2 hour or more, depending upon the amount of moisture in the system.

A better alternative is to install a filter drier in the system prior to evacuation. Use a size which is rated for the BTU or tonnage capacity of the unit. The filter drier will ensure that any moisture left in the system after evacuation will be trapped where it cannot cause ice formation.

Blocked capillary lines sometimes may be cleared by the use of a capillary tube chaser kit, which contains several spools of lead wire designed to fit into popular sizes of capillary tubes. A hydraulic pump, hand operated and capable of exerting up to 5000 PSI, is used to push a small length (3/8 inch or 1 cm) of lead wire through the capillary tube to act as a piston and clear the obstruction. The lead wire then falls into the evaporator coil where it remains harmlessly.

A capillary tube gauging tool supplied with the kit allows the service technician to determine the size of the capillary tube. This allows the selection of the correct size lead wire, and is also useful to help select the correct size tubing when replacement is necessary.

Capillary tubes should never be cut using a wheel type tube cutting tool, since the action of the tool will distort the small size inner diameter of the tubing. Instead, an ordinary file can be used to score one side of the tubing so that it can be bent back and forth until it breaks. The file can then be used to remove any burrs and smooth the outer periphery of the tubing. It is important that no metal chips or other debris be allowed to enter the tubing.

When replacing capillary tubes, it is important to use the exact size and length of the original. These parameters determine the restriction to the liquid refrigerant, and are selected by the manufacturer of the unit for optimum performance.

Before assembling a capillary tube to the evaporator or condenser section, the end of the tube should first be checked to ascertain that the small opening is clear and not obstructed in any way. The larger tube may be crimped with a Vise Grip® or similar type pliers so that it hugs the capillary tightly. The excess periphery of the larger tube should be squeezed shut so that a bead of brazing filler material can be applied during the brazing operation. Always insert the capillary tube at least 3/4 inch (2 cm) into the larger pipe before crimping and brazing, to ensure that none of the liquid solder will flow to the end of the tube and clog it.

5.20 Expansion Valve Service

Improper operation of an expansion valve will result in insufficient or no cooling, or ice buildup on the evaporator coils or suction line. System pressures (both high side and low side) will be below normal when the expansion needle valve is stuck in a closed position, and above normal when stuck open.

Before condemning the expansion valve as defective, the service technician should check the sight glass to verify that the system is sufficiently charged with refrigerant. Bear in mind that a defective expansion valve may prevent a normal condition at the sight glass. The manifold gauge set should be connected to the system to monitor pressures. If the expansion valve is equipped with a sensing bulb, it should be checked to be certain that it is securely clamped to the evaporator outlet pipe. These bulbs may also be covered with thermal insulation that should be intact.

Normal heat load, in the form of proper air flow, must be placed on the evaporator coil to prevent freeze-up. The evaporator blower section should be checked for proper line voltage, tight belts (if used), and clean air filters. In three phase systems the direction of rotation of the blower motor should be checked, since a reversed phase sequence will cause reverse motor rotation. A properly operating air conditioning system requires at least 400 cubic feet per minute air flow through the evaporator coil assembly. Evaporator air inlet temperature during the checkout should be at least 75 degrees F (24 degrees C) to ensure sufficient heat load on the evaporator.

Expansion valves can fail in either an open or closed position, or be stuck anywhere in between. Some valves are equipped with filter screens that can become clogged, simulating a valve stuck in the closed position. Service of expansion valves will require the system to be discharged of refrigerant. If available, the use of a recovery system to collect the refrigerant during the discharge procedure is recommended.

As a last resort, field adjustable expansion valves should be checked for proper adjustment before a decision is made to discharge the system and replace the expansion valve. This is accomplished by checking the system for proper superheat, which is the temperature difference between the expansion valve sensing bulb and boiling temperature of the liquid refrigerant in the evaporator.

A clamp-on thermometer may be used to measure evaporator return line temperature at the point where the sensing bulb is mounted. With the system running, the low-pressure manifold gauge temperature scale (for the system refrigerant) will indicate the temperature of the boiling refrigerant. The difference between the two temperature readings is the superheat.

If the system manufacturer's service manual is available, it may specify the proper superheat. Otherwise, 10 degrees F (5.6 degrees C) is a good starting point. Adjustment of superheat values much over this amount may cause a starved evaporator and reduced cooling capacity.

If normal cooling cannot be restored by adjustment of the expansion valve, and all other possible sources of trouble have been eliminated, the expansion valve can be removed from the system and checked on the test bench as illustrated in Figure 5.17. If equipped with a filter screen, this may be removed for cleaning or replacement, if necessary.

Compressed air or nitrogen may be used for this test. The thermostatic expansion valve is checked for operation by inserting the sensing bulb into the bottom of a container which is completely filled with finely cracked ice, so that bulb temperature is maintained at 32 degrees F (0 degrees C). The adjustable needle valve is set to allow a very slight bleed of gas through the test setup. The pressure read by the gauge should indicate a superheat of about 10 degrees F (5.6 degrees C). For R-22 this pressure is 45 PSI, and for R-12 it is 22 PSI. Adjustable expansion valves may be set to reflect the proper superheat. When the bulb is removed from the ice mixture and warmed by hand, the expansion valve should open smoothly and permit a full flow of air or gas, as indicated by an increase in pressure reading.

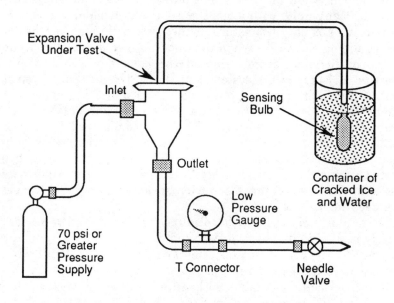

Figure 5.17   Test Setup for Thermostatic Expansion Valve

5.21 Filter/Drier

Filter/driers are cylindrical assemblies which contain a moisture absorbing material, called a desiccant, and a fine mesh filter screen. They are designed to perform two essential functions in an air conditioning system: trapping any possible moisture or other contaminants and preventing small particles, as small as 5 microns, from passing through the system.

A filter/drier is often used in both the suction line and liquid line of an air conditioning system. The suction line filter prevents particles from entering the compressor inlet, and the liquid line filter keeps the capillary tube or expansion valve clean. Heat pump systems require the use of bidirectional components since refrigerant flows in both directions.

Liquid line filter/driers may also include a sight glass to allow the service technician to monitor the flow of refrigerant and determine if the system is low on charge as evidenced by a steady flow of gas bubbles. Many units with a sight glass also provide an indication of the relative amount of moisture in the system through the use of a color changing chemical. When the color is green or blue, the system has no or little moisture. Should the color turn pink, there is excessive moisture in the system that should be removed by replacement of the filter/drier and total evacuation and charging.

Some filter/driers may have two access ports that permit measurement of its pressure drop. This is accomplished by taking a reading at each port and subtracting the results. The pressure drop is an indication of the level of restriction due to its filter screen being covered with debris, and indicates when replacement is necessary. The manufacturer of the filter/drier specifies the allowable pressure drop across the unit.

The amount of water absorption capacity of a filter/drier is directly related to its physical size, and the type of refrigerant in the system. Table 5.4 illustrates the capacity of typical filter/driers used in air conditioning service.

Filter/driers are available with either solder fittings or flared connections. They are rated by BTU capacity or tonnage of the air conditioning system, and may have different ratings in accordance with the type of refrigerant used.

When a system experiences a burned out compressor, a suction line filter/drier should be installed after purging to "clean up" the system and prevent contaminants from entering the new compressor.

Table 5.4   Typical Chart of Liquid Line Filter/Driers
Showing Water Absorption Capability

Desiccant Volume in Cubic Inches	Water Capacity in Drops (R-22)	Water Capacity in Drops (R-12)
3	31	34
5	66	72
8	107	117
16	190	206
30	383	416
41	562	611

By checking the pressure drop across the filter/drier after the system has been in operation for some time, any possible blockage of refrigerant flow can be determined, and the filter/drier replaced if necessary.

Air conditioning systems that have been exposed to the atmosphere for a substantial length of time may collect enormous amounts of moisture. One way to facilitate moisture removal is to install and replace as many filter/driers as required, one at a time, to trap the water and remove it from the system. Each time a new unit is installed, the system should be evacuated and charged (at least partially) and operated for sufficient time to allow the drier to soak up the moisture passing through.

Small window air conditioning units are usually not supplied with filter/driers, but they may contain small filter assemblies to trap particles. It is usually not necessary to add a filter/ drier to the system when a compressor is replaced, but it should be purged with liquid R-11 or another suitable product if the compressor has failed dirty.

5.22 Silver Brazing Alloys

Silver brazing alloys are available from several manufacturers, one of which is Lucas-Milhaupt, 5656 S. Pennsylvania Avenue, Cudahy, WI 53110. They supply many different alloys of silver brazing materials, some of which are designed for air conditioning and refrigeration use.

Silver brazing alloys, called filler metals, are composed of silver, copper, and zinc, and may also contain other metals or elements such as cadmium. A disadvantage of using any alloy that contains cadmium is that at brazing temperatures, cadmium will volatilize and produce toxic fumes. For this reason, cadmium-free alloys are recommended.

Table 5.5 Typical Silver Brazing Alloy Filler Materials

Type	Melting Temperature	Composition (% WT.)
Easy-Flo35	1295 F 702 C	35 Ag 26 Cu 21 Zn 18 Cd
East-Flo 45	1145 F 618 C	45 Ag 15 Cu 16 Zn 24 Cd
Sil Fos	1300 F 704 C	15 Ag 80 Cu 5 P
Sil Fos 2	1325 F 718 C	2 Ag 91 Cu 7 P

The brazing operation should always be performed in a well ventilated area. There is always the possibility of dangerous fumes being generated from fluorides in fluxes or coatings on base metals such as zinc or cadmium. The use of ventilating fans or exhaust hoods is a good practice.

A group of alloys illustrated in Table 5.5, manufactured by Lucas-Milhaupt, Inc., is recommended for copper-to-copper brazing. This manufacturer, and others, produce many other alloys which may also be used for refrigeration and air conditioning work. The supply house that stocks brazing supplies will be able to recommend various filler materials and fluxes that may be used.

5.23 Brazing

Brazing is the joining of metals through the use of a filler material whose melting temperature is below the melting point of the metals being joined. Unlike soldering, brazing is performed at much higher temperatures (above 840 degrees F or 449 degrees C). Since filler materials used in air conditioning work are usually an alloy of silver with other metals, the correct term for this process is silver brazing.

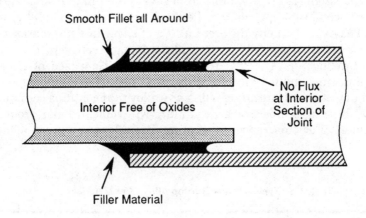

Figure 5.18   Cut-away View of Properly Brazed Joint

Brazing is the preferred method of joining refrigerant lines and components because of its tensile strength and ability to withstand shock and vibration. Such joints are generally easy to make and economical. The necessary skill to produce good brazing joints is easily acquired. Figure 5.18 illustrates a cut-away view of a properly brazed joint.

Although the brazed joint will "make itself" through capillary action, it is up to the operator to provide the proper conditions and follow the prescribed procedures. These can be broken down to six basic steps.

1. The parts to be brazed together must be clean. Any contamination (scale, oil, etc.) will interfere with the brazing process and prevent capillary action from drawing the filler material into the joint. Any part of the metal surface that does not "wet" properly by the action of the flux and filler metal will produce a void. Such a condition can allow refrigerant leakage in an air conditioning system, and the joint will have to be redone properly.

   Parts should be cleaned of oil or grease first, using an approved degreaser. Any residual coating such as scale or oxide on the metal surfaces can then be cleaned using fine emery or steel wool. It is important that care be taken to prevent any particles from entering the interior of the parts being cleaned. Dirt is an enemy of any air conditioning system and will cause problems and possible damage to working parts.

2. The fit between the parts to be joined should be reasonably good. Capillary action depends upon the two surfaces that are joined to be within close proximity. The fit does not have to be precise; an easy slip fit will produce an excellent brazed joint.

   Some brazing materials, such as Easy-Flo 35, are designed for joints which have more clearance than normal. Such a joint will be very strong, but it will not have the tensile strength of one which is produced with close-fitting parts.

3. Flux is essential to the brazing process. It coats the metal surfaces and prevents the formation of oxides that prevent wetting and bonding of the parts. Flux is capable of dissolving and absorbing any residual oxide that is left on the metal surfaces after the cleaning operation.

Table 5.6   Four Stages of Flux during the Brazing Operation

Temperature	Appearance of Flux
212 F 100 C	Water boils off, leaving a flux residue.
600 F 315 C	Flux becomes white and puffy as it starts to work.
800 F 425 C	Flux turns milky and lies against the work.
1100 F 595 C	Flux turns clear and is active. Bright metal surface is visible underneath. Brazing material may be applied to the joint as heat is applied to form the brazed joint.

Brazing flux is a water soluble compound that may be applied to both surfaces of the parts to be joined with a brush or any other suitable instrument. It may be thinned with water to obtain the desired consistency, and should be generously applied to the joint just prior to the brazing operation. Care should be taken to ensure that no flux material enters the interior of the air conditioning system.

There are different formulations for flux, depending upon the application, but a general-purpose flux will be a good choice for air conditioning work. The supplier of brazing materials can provide information on the best flux to use.

The coating of flux on the parts will act as a temperature indicator that will inform the service technician of the progress of the brazing operation. There are four distinct stages of flux condition when heated, illustrated in Table 5.6.

4.   The parts to be brazed should be self-supporting during the brazing procedure. This will prevent movement during the operation, which can result in a defective joint. If possible, position the parts to be joined so that the force of gravity holds them in place. In a difficult situation, clamps may have to be used to support the parts. Care should be taken that such clamps are not overheated during the brazing process.

5. The brazing operation is not complicated. It consists of heating the metal parts to be joined and applying the filler material so that it may flow through the joint. An oxy-acetylene, air-acetylene, or Mapp gas torch are good choices for the brazing procedure.

   Oxides should not be allowed to form on the interior of tubing or fittings during the brazing operation, as will happen if the system contains air. This is accomplished by providing a minute flow of an inert gas, such as nitrogen, throughout the system during the brazing process, and until the parts have cooled. Failure to prevent the formation of oxide can cause problems with small orifices or moving parts as the oxides are carried through the system along with the refrigerant.

   Service tubes on compressors are brazed at the factory to the compressor case with similar brazing materials. For this reason the brazing torch should be kept away from such joints to prevent melting and loss of mechanical attachment of the parts to the compressor shell.

   Both parts that are to be joined are heated with the torch. The appearance of the flux will indicate when the parts are hot enough to melt the filler material, which occurs when the flux is clear and active. Thicker parts will require more heat than thinner ones, and some of the heat will be drawn away through conduction of the metal parts.

   The brazing material should be fed to the joint where the parts are joined so that it will melt and be drawn in by capillary action. Filler material tends to flow towards areas of higher temperature. For this reason the torch should be applied to the opposite side of the joint from the point where the filler material is being applied, so the heat will draw the material throughout the joint. A smooth fillet appearing at the joint indicates that sufficient filler material has been applied.

   When the brazing operation is completed, it may be allowed to cool naturally, or the process may be hastened by carefully quenching the assembly with a rag that has been soaked in hot water.

6. The final step in the brazing process is to clean the joint of all residue and flux. This material is chemically corrosive and may weaken a joint in time. A coating of flux may also mask a hidden void that will eventually leak refrigerant.

Flux removal is usually accomplished using water. Stubborn residues may be cleaned off with a fine wire brush, which should flake off easily, taking care to avoid damage to thin metal tubing. Any residue which is difficult to remove may be saturated with oxides, or may be due to overheating of the parts. After removal, the joint should be inspected for any voids or sections where the parts did not properly wet during the brazing process.

When the brazing operation is completed, the air conditioning system must be pressurized with refrigerant, and a leak detector used to determine if the parts are leak tight. If this is not done, all refrigerant can be lost through a leaking joint.

5.24 Leak Repair

An air conditioning system which is low on refrigerant charge most likely will have a leak which can be as small as less than 1 ounce per year to a catastrophic condition in which the total refrigerant charge is dissipated in minutes. Most leaks will fall somewhere between these two extremes.

Low refrigerant charge can be detected by several means, such as frost forming on the evaporator coils, insufficient line current being drawn by the unit (compared to its nameplate value), or a steady flow of bubbles in the liquid line sight glass if so equipped. Additional symptoms of low refrigerant charge are an evaporator which is only partially cold or condenser air flow which is relatively cool, as compared to other air conditioning units of similar capacity.

Refrigerant will leak out of any part of the system which has deteriorated due to age, vibration, defective construction, or corrosion. In particular the most likely places where leaks occur is in the various brazed or threaded connections between components. Additionally, fatigue of any of the metal parts of the sealed system will eventually result in refrigerant leakage. Common reasons for leakage are fatigue caused by excessive vibration, and perforation of tubing caused by the rubbing of one part on another.

Very small leaks, such as a few ounces a year, are sometimes difficult to find, especially on systems which have large-size evaporators and condensers. Larger leaks can sometimes be spotted by a simple visual inspection of the entire unit. Many times such a leak will be evidenced by an oil stain surrounding the leak, which is caused by the system oil leaking out with the refrigerant.

It cannot be overemphasized that inspection of the unit for oil stains can go a long way in reducing service time. When such a stain is located, a halide or electronic leak detector can often pinpoint the exact location of the leak. Larger leaks can be seen by using a soap solution which will provide visual indication of the location and magnitude of the leak.

If it has been determined that a faulty brazed joint is the source of trouble, the only way to properly repair the connection is to separate the parts, examine them for corrosion or weak sections, and thoroughly clean them before reassembling. If the metal is excessively corroded, it may not be possible to obtain a solid leak-tight joint. A better way to repair the leak is to cut away the corroded parts and replace them with new tubing or fittings. Although this may require one additional solder connection, it is far better to do the job properly the first time rather than attempt to repair a badly corroded joint.

Leaks that are caused by metal fatigue that result in a crack in a pipe can sometimes be repaired by flowing silver brazing material over the affected area. The section of pipe to be repaired should be thoroughly cleaned and examined for weak spots. If necessary, copper tubing with a large opening can be successfully repaired by forming a small piece of sheet copper over the opening and brazing the parts together.

When the repair has been made, the mechanical problem which caused the failure must be corrected to prevent it from happening again. For example, if the failure was due to interference between a copper tube and some other part of the unit, the tubing should be gently reformed to eliminate the interference.

If the metal parts have failed due to excessive vibration, the source of the problem should be corrected if possible, and the metal parts secured so that they do not vibrate when the unit is in operation.

If an extreme case of metal corrosion is encountered, such as in the bottom coils of the condenser which have rested against the steel tray of a very old unit, it may be possible to place the unit back in service without replacement of the entire component by modifying the coil so that the defective section is eliminated from the system. Such modification will eliminate a small portion of the coil assembly and will have some effect on system operation, but it often can be used to save a unit which otherwise might have to be scrapped.

Repairing leaks in units which employ aluminum tubing can be difficult. Although there are many aluminum solders and epoxies on the market, the service technician must be proficient in using such materials to be successful. The melting point of aluminum tubing is about 1200 degrees F (649 degrees C) and the melting point of the solder may be 700 degrees F (371 degrees C). This necessitates gentle heating of the parts so that the melting point of the solder is reached without approaching the melting point of aluminum. Such tubing is very thin and can easily be destroyed by excessive heat.

It is suggested that before attempting an aluminum repair, practice on scrap aluminum tubing, evaporators, or condensers until the necessary skills have been attained.

5.25 Condenser and Evaporator Service

Condensers and evaporators are passive devices which rarely require service except for periodic cleaning to remove accumulated dust and debris on the fins. Such accumulated material will hamper operation of the unit, resulting in less cooling and lower efficiency. Most air conditioners should be inspected and cleaned at least annually to remove any dirt buildup. If a unit is operated under severe conditions, or without proper filters, cleaning will be required more often.

Suppliers to the air conditioning service industry provide a multitude of cleaning materials and equipment which can convert a greasy, grimy condenser coil back to like-new condition. These devices use special chemicals and detergents which are sprayed on the coils at high pressure, and have been formulated to clean condensers without damage.

When using such products, be sure to disconnect electrical power to the unit before starting the cleaning procedure. It is extremely important that any electrical or other sensitive components within the unit are not inadvertently sprayed. To do so may result in permanent damage to the parts. If necessary, cover such parts before spraying.

Window-type air conditioners are especially susceptible to build-up of a thick layer of dirt on the evaporator coils, since they are often operated with extremely dirty filters, or none at all. These coils can be successfully cleaned by thoroughly coating the evaporator fins with a common household spray cleaner such as fantastik® and using the stream of a common garden hose to flush the dirt away with the detergent.

This method of cleaning the evaporator coils is relatively safe, since the water spray is contained within the evaporator section of the unit. It therefore will not affect either the motor or electrical components which are placed behind partitions. When the cleaning process is completed, however, the unit should be allowed to dry thoroughly before applying power.

Using a household cleaner on the condenser coils in a similar manner as with the evaporator may also be done, but in this case the electrical parts of the unit will be exposed to the water spray. Be sure to cover such parts before the cleaning process. Note that household cleaners are not strong enough to thoroughly clean a very dirty or greasy condenser coil. For this purpose the stronger chemical cleaners designed for that purpose can be used.

Mechanical failure of either the evaporator or condenser coil assembly is relatively rare. However, as with any passive device that operates under pressure, leaks can occur. Additionally, these components are subject to damage from vibration and abuse, due to their fragile nature and exposed location. When such damage occurs the refrigerant leak must be located and repaired using traditional service techniques. If the failure was due to excessive vibration, its cause should be corrected.

Excessive damage may result in an evaporator or condenser coil which is beyond repair. Replacement assemblies are usually available from the manufacturer of the unit, or a suitable repair may be accomplished by salvaging a serviceable part from a discarded air conditioner unit. When substituting parts, be sure to count the number of coils of the original unit and use a replacement which has a similar number of coils and is about the same size in area.

5.26 Fault Analysis

Air conditioners are, for the most part, not complicated devices and it is often possible to make a quick analysis of a fault in the system by operating the unit and gathering as much information from the owner as to the nature of the problem. In many cases fault analysis can be accomplished using sight, sound, and feel, and without the help of any instruments or tools whatsoever.

To accomplish a reasonable quick fault analysis, the source of the problem is determined by an orderly and logical sequence of checks. When a certain section of the unit appears to be at fault, the appropriate instruments (manifold gauge set, voltmeter, ammeter, etc.) may be connected to the system to verify the suspected fault. Some problems may be caused by different components in the system even though the symptoms are the same.

The best possible aid to troubleshooting an air conditioning system is the service manual which may be supplied by the manufacturer of the unit. Even small air conditioners, such as window units, are often shipped with some documentation which provides pertinent information.   Central and commercial systems are usually sufficiently complex to require a schematic or wiring diagram for troubleshooting electrical problems, and should be obtained if possible.

The following diagnostic chart, Table 5.7, is provided to help in locating the area of fault. To isolate the problem further, refer to the previous sections which discuss the parts, operation, and repair in detail.

Table 5.7 Troubleshooting Chart

Symptom	Possible Cause
Compressor will not run; no line current draw	No power to unit. Overload or pressure control switch open.
Compressor trips overload or circuit breaker	Defective capacitor; compressor winding open or shorted; internal mechanical problem; unequalized system pressures.
Compressor short cycles	Defective overload; excessive head pressure, low system refrigerant charge, intermittent control. Thermostat differential too narrow.
Compressor runs; condenser cold	No system refrigerant; restriction in system; compressor defective.
Compressor runs; no cooling	Loss of system refrigerant; restriction in line; defective expansion valve; mechanical defect in compressor.
Frost on evaporator	Low system charge; stuck or defective expansion valve; restriction in line; insufficient heat load on evaporator.
Excessive discharge pressure	System overcharge; insufficient air flow through condenser; air in system.
High suction pressure	System overcharge, defective expansion valve; defective compressor; excessive heat load on evaporator.
Liquid line cold	Restriction in filter/drier
Noisy operation	Compressor mechanical problems; defective or dry motor bearings; fan blade interference.

Evaporative Coolers

6.1 General Information

In areas of the world where humidity levels are naturally low, such as in the southwest, an evaporative cooler (sometimes referred to as a swamp cooler) actually does a better job of cooling and conditioning the air than does a refrigeration system. This is because refrigeration units remove moisture from the air, and special humidifiers have to be installed to restore the moisture balance. Evaporative coolers, on the other hand, add moisture to the room during the cooling process. As a bonus, evaporative coolers are cheaper to install, operate, and maintain.

The evaporative cooler operates on the principle of an evaporating liquid (water in this case) which removes heat from the surrounding air. The effect is the same as if a small amount of rubbing were placed onto the skin. The cooling effect due to the evaporation of the alcohol is immediately felt.

6.2 System Operation

In the evaporative cooler, air is forced through coarsely-woven cloth, called an evaporation pad, that is saturated with water. Through natural heat transfer, some water molecules will spontaneously go from a liquid state to the vapor state without the application of boiling heat. The phenomenon is called evaporation.

In the process of evaporation, the vaporizing molecules must absorb enough heat energy from the surrounding air to satisfy the law of thermodynamics. The principle is identical to that used to cool air that is passed over a block of ice to cool a room. In other words, 144 BTUs of heat are extracted from the air passing through the pad for every pound of water that changes from a liquid to a vapor through evaporation. This is what causes the cooling effect.

However, evaporation coolers only work when the humidity is low enough to permit a noticeable temperature differential. As the humidity rises, fewer and fewer water molecules can vaporize because the air is already heavy with water vapor and there is little room available for newcomers.

As a rule, the average humidity should be no more than 40% for effective evaporative cooling. For example, the best you can get using evaporative cooling on 90 degree F (32 degree C) air with 67% humidity (a humidity many "reportedly arid" regions exceed during certain parts of the summer because of thunderstorm activity) is less than 10 degrees F (5.5 degrees C).

6.3 Rating An Evaporative Cooler

Like refrigeration systems, evaporative coolers are rated in tons or BTUs per hour. However, the same method of matching the BTU rating to the building load doesn't apply because the load rating is measured at a specific temperature and humidity, usually 85 degrees F at 30% humidity. As the two fluctuate, so does the BTU rating.

A better method to calculate the required size of an evaporation cooler is to design for a given differential between outside and inside temperatures. For example if the normal daytime summertime ambient temperature is usually about 90 degrees F (32 degrees C), a system designed for 15 degrees F differential will be able to maintain 75 degrees F (24 degrees C) inside.

A psychrometer may be used to determine the temperature differential between outside air and inside air. A psychrometer is an instrument which contains two thermometers, one with a moistened bulb and one with a dry bulb, that are used in conjunction with a psychrometric chart to determine relative humidity level (see Chapter 2). Because of evaporation, the wet bulb will always read less than the dry bulb (except at 100% relative humidity), which refrigeration technicians use to determine the relative humidity using a conversion chart.

When the psychrometer is used with evaporation coolers, the temperature difference between the two thermometer readings indicates how many degrees of cooling can be expected for that climate. If the psychrometric reading is taken during the heat of the day, as it should, the lower reading will also indicate the room temperature the evaporation cooler will maintain.

Next calculate the amount of air flow needed to maintain the temperature differential for the volume of the room or building. As a rule of thumb, the air should be exchanged at least 5 times an hour (once every 12 minutes) for even temperature control.

A simple way to calculate the amount of air flow needed is to divide the cubic volume of the room (in feet) by the number of minutes you want the air exchanged. For example, a standard 8 by 12 room contains 768 cubic feet of air. To move that much air every 12 minutes you need a 64 CFM (cubic feet per minute) blower. Of course, doors and walls inhibit the flow of air, a factor which must be taken into account. A typical one-story house with 1200 feet of floor space needs about 2500 CFM of air flow.

When recommending an evaporation cooling unit, keep in mind that the peak—not the average—cooling requirements are what determine the size of the cooler. Unlike refrigeration systems, which cycle off and on to maintain a constant temperature, the evaporation cooler runs 100% of the time. Installing a unit that only meets average requirements (as is recommended for refrigeration cooling) puts the room temperature on a roller coaster ride that follows the outside temperature. Installing an oversized unit eliminates the problem.

Instead the room temperature is controlled by adjusting the flow of air through the evaporation pads. Lowering the speed of the blower motor lowers the amount of air drawn through the pads and the amount of water being evaporated, which in turn lowers the amount of BTUs extracted from the outside air. Most evaporative coolers have a three-speed motor and manual speed switch for that purpose.

6.4 Evaporation Cooler Construction

Evaporative coolers range in size from small window units to roof-top boxes to large industrial towers. Portable tabletop models were once the rage of apartment dwellers, but they have all but disappeared from existence.

All evaporative coolers are constructed similarly and differ little in appearance other than size. Basically, the cooler consists of a square metal box with four open sides to which the evaporative pads are affixed. The pads are kept moist by an internal water pump. A blower or fan is placed in the middle of the metal frame to draw air from the outside through the damp pads and into the building.

6.5 Blower Motor

Central to an evaporative cooler is a squirrel cage blower or a fan used to suck the outside air through the evaporation pads. Blowers are usually found in downdraft roof top models where the blower mounts to the floor of the cooler box and forces the cooled air out the bottom of the cooler. Fans are most often found in window mounted units and face to one side of the box, forcing cooled air into the room through the attached window opening.

The squirrel cage blower is usually belt driven rather than direct driven. The motor is of the general-purpose utility type with a 1/4 to 1/2 horsepower rating. Most often the motor uses a starting relay.

The starting relay contains an armature which is actuated by the heavy inrush current of the run winding when power is first applied. This connects the starting winding to the power line and causes a tremendous starting torque to be developed. When the blower is up to speed, the running winding current quickly drops to its normal running current. This causes the relay to disengage, opening its contacts and disconnecting the starting winding from the circuit.

Most motors are equipped with an overload cutout, which provides protection in the event that the motor doesn't come up to speed within a given amount of time after power is applied. The starting current is about 10 times that of running current, and can quickly burn out the winding out if sustained for more than a couple seconds. The overload sensor may be part of the relay or a thermal switch built into the motor. In either case, the cutout generally recycles itself within a couple minutes. However, if the problem is not located and corrected within a few attempts at starting, the overload sensor and/or starting winding will be permanently damaged.

Very large commercial evaporation coolers and towers have blower motors that are rated at 5 horsepower and more. Motors with ratings between 1/2 horsepower and 5 horsepower normally use starting capacitors to give the kick needed to get the blower up and running. The speed of these motors may be controlled using a manual switch. Blowers that need more than 5 horsepower of drive very often use large synchronous motors (very large clock motors) which run at a constant speed.

6.6 Water Pump

It is the job of the water pump to keep the evaporation pads wet during operation. It does this by supplying a constant flow of water that trickles through the pad from top to bottom.

The water pump, which looks every bit like a basement or sump pump, is a very inexpensive siphon pump that stands upright in a basin of water, called the catch basin, at the bottom of the cooler. The pump is located in a plastic mesh basket that filters out large rust and dirt particles that can clog the arteries of the water system.

The water is pumped from the catch basin up to the top of the cooler where it is distributed through tubing and troughs to the evaporation pads mounted vertically along the outside walls of the shell.

The motor part of the pump is above water and may or may not be sealed against moisture. Most pumps of this type use universal AC/DC motors (like drill motors) to spin the impeller in the base of the pump assembly. While it is possible to replace the brushes should they wear out, it is more likely the motor bearing or impeller will succumb to the effects of water corrosion long before the brushes need replacing.

Replacing the water pump is as simple as lifting the old pump out of its stainer basket and putting the new pump in its place. Only two electrical connections are needed, and they can be made using water-proof wire nuts to the old pump's power cord.

6.7 Float Valve

The level of the water in the catch basin is maintained by a cutoff float valve. The arrangement of the valve assembly is identical to that used in the tank of a toilet to keep it from overflowing.

Adjustment of the water height is done through the usual trial-and-error method of bending the float rod until the valve cuts off at the right level. As the valve ages and the seat becomes more compressed through constant use, the float level will need to be adjusted downward until a point is reached where the valve no longer seals properly, in which case the only remedy is replacement.

Replacing the valve usually requires the purchase of a float level assembly. The only attachment the valve assembly makes to the cooler is through a small hole in the catch basin area of the metal housing. Replacement requires two wrenches and possibly a pipe flaring tool.

6.8 Evaporation Pads

Last but not least are the evaporation pads. These are woven pads of a coarse grass that readily absorb and hold water.

The grass pads come encased in a plastic netting that holds the pad in shape during handling. The pads are mounted in the metal pad frames that make up the sides of the cooler housing. Built into the top of the pad frame is a water distribution trough with drip slots spaced at regular intervals so that the water delivered from the water pump evenly soaks the pads with water.

6.9 Annual Inspection and Maintenance

Evaporation coolers require a schedule of yearly maintenance that is determined by the seasons. At the first signs of warm weather, an inspection of the evaporation cooler should be made. A checklist is provided in Figure 6.1 for your convenience.

Remove old evaporation pads   _____
Inspect pad frames for rust and calcium   _____
Lubricate blower or fan motor   _____
Inspect/replace blower belt   _____
Clean catch basin   _____
Inspect/adjust float level   _____
Inspect water pump   _____
Install new evaporation pads   _____
Reassemble   _____

Figure 6.1 Annual Inspection and Maintenance Checklist

Evaporation Pads

The evaporation pads should be replaced annually, preferably at the beginning of the summer season. Pads come in standard sizes of 24 and 36 inches square. Smaller sized pads are usually increments of these two sizes, and one 36 inch pad can be cut in half for use as two pads in an 18 inch cooler.

To replace a pad, remove the pad frame from the cooler by twisting the plastic retainer clips so that the frame can be pulled forward. Lift the frame up and pull out for removal.

The evaporation pads are held in place with a wire retainer located on the inside of the pad frame that snaps into small holes located along the edges of the pad frame. Bend the wire retainer slightly so that it gives enough to clear the mounting holes. Remove the old pad.

Repeat for all sides, carefully noting which frame came from which side of the cooler. Before installing the new pads, continue with the maintenance checklist.

Evaporation Pad Frame

The evaporation pad frame should be inspected for rust and calcium deposits. Pay special attention to the drip slots in the water trough at the top of the frame.

Often these drip slots become clogged with calcium deposits. If the deposits are light, the slots can be cleared using a wire brush and utility knife. Heavy calcium buildup requires the use of a chemical solvent, such as Lime Away or Tilex. Calcium deposits that accumulate on the ventilation fins of the frame can be removed in the same way.

Rust spots should be scrubbed with a wire brush to remove loose rust particles then treated with a coat of rust inhibiting primer paint, such as Krylon red primer in the spray can. After the primer is dry, a coat of color paint may be applied, but it is not necessary unless the primer can be seen from the outside of the cooler.

Blower Motor

Inspect the blower motor belt for wear. Replace if cracked or worn. New belts can be obtained from most auto supply stores.

Clean and lubricate the blower motor or fan according to the manufacturer's instructions. Check that the shaft spins freely. Prolonged periods of non-use can cause the shaft to stick, which can usually be broken free by hand.

When sealed bearing motors fail because of dry or frozen bearings, it may be possible for the service technician to coax some oil into the bearings along the shaft, thus restoring lubrication and adding service life to the motor. Non-detergent SAE 20 weight oil should be used. Sometimes complete disassembly of the motor will permit the bearings to be properly lubricated. If you proceed this far, drill an outside oiling hole and install a plastic oiler tube for future lubrication. If attempts at lubrication do not succeed, replacement is necessary.

Burnout of the starter winding is not uncommon and should be suspected if the starter relay frequently or constantly kicks out after applying power. A burned out or damaged starter winding may be detected by a check of the resistance of the winding after disconnection of the leads from the electrical circuit. Burned out motors cannot be repaired, and replacement is required.

Catch Basin

Remove all loose debris from the catch basin. Inspect the pan for signs of rust and water leakage. Treat the rust with a wire brush and rust inhibiting primer as outlined above.

Water leaks are caused by rust spots that have etched through the metal. If not too severe, the leak can be stopped by cleaning the rusted area thoroughly and covering it with a good waterproof roofing patch, such as Henrys.

Many coolers come with their catch basin already protected by a waterproof coating. But through use the coating may develop cracks that let water get to the metal underneath. Rust that develops from leaks of this kind are very hard to spot, and usually go undetected until too late to fix. If a spot of a coated pan looks suspicious, don't take chances. Apply a liberal coat of a roofing sealant over the area.

Float Level

Turn on the water and allow the catch basin to fill. Check the operation of the water valve by lifting the float until the water turns off. Do not force the float; doing so can rupture or distort the valve seat. If the water flow doesn't completely stop, replace the valve.

Check the level of the water as the basic fills. If necessary, adjust the float level by bending the rod until the proper height is achieved. Proper water level is when the bottom of the water pump is completely covered by an inch of water.

Water Pump

Turn on the water pump. BE AWARE that doing so will also turn on the blower or fan, so keep your hands and clothing clear of all moving parts.

Make sure the pump is pumping water to all the water feeder outlets located at the top of where the evaporation pad mounts. A steady stream of water should be seen in the middle of each window if working properly.

If no water flow is visible, turn off the cooler and remove the drive belt from the blower (or disconnect the fan) for safety purposes because you will now have to put your hands inside the cooler. With your fingers, see if the shaft to the impeller spins freely. If may have become stuck during storage, and your finger pressure may break it free.

If the pump shaft spins freely, suspect a clogged line. The water pump is connected to the evaporation pad water feeder lines by a rubber hose that may be removed. With power applied, a strong stream of water should gush from the rubber hose. If not, replace the pump.

If the pump is working, one or more of the feeder lines is clogged. A quick way to clear a clogged line is to run a piano or guitar wire through the tubing, knocking out the accumulated debris. Flush the tubing well with water as you proceed.

Install The New Pads

Finally, install the new evaporation pads and replace the pad frames in the cooler. Trimming of the pads shouldn't be necessary unless the pad is a multiple of a larger size pads. If necessary, cram the pad into the corners with your fingers. Neatness doesn't count because the pad will change size and shape once it becomes saturated with water.

6.10 Winterization

Unlike refrigeration systems, evaporation coolers have to be prepared for the winter. The procedure is simple, and only takes a few minutes.

First turn off the water at the source and drain the catch basin. Most coolers have a plug located near the water valve for this purpose. Window mounted units may have the drain plug on the bottom of the cooler. Never let the water set in the catch basin to evaporate on its own; that's how rust starts.

Above all, never leave the water source turned on during the winter. If you do, there is a good chance the pipe will freeze and burst, forcing you to turn off the water anyway and replace the pipe come spring.

Cover the cooler with a hood. The hood prevents warm air from inside the house from escaping to the outside, thus lowering your winter heating bills. Cooler hoods are sold by most hardware stores, or are available from the manufacturer. Alternatively, the cooler may be wrapped with heavy sheet plastic.

Automotive Air Conditioning

7.1 General Information

Most automobiles and other vehicles manufactured today are equipped with factory installed air conditioning systems. This trend has been in existence for over ten years, which means there are literally hundreds of millions of vehicles on the road in the United States that contain air conditioning systems. All of these systems will require service during their normal lifetime of ten years or more. Although there are hundreds of different vehicle models, both domestic and foreign, all A/C systems are very similar. These automotive air conditioners operate under severe conditions and have unique service requirements.

Vehicular air conditioning systems are similar in some ways to those which are found in homes and offices, but there are differences in the components and controls which are necessary due to the unique operating requirements. These differences are not overly complicated, but must be properly understood for proper diagnostic and repair procedures. Anyone competent in the repair of hermetic air conditioning systems should be able to diagnose, troubleshoot, and repair automotive air conditioning systems. Figure 7.1 illustrates the makeup of a typical automotive air conditioning system.

Automotive air conditioning systems are equipped with both low-side and high-side service connections, most of which contain spring-loaded valves. Some manufacturers provide different sized fittings for each to prevent accidental misconnections. The low-pressure connection is the common Schrader fitting which matches the charging manifold hoses.

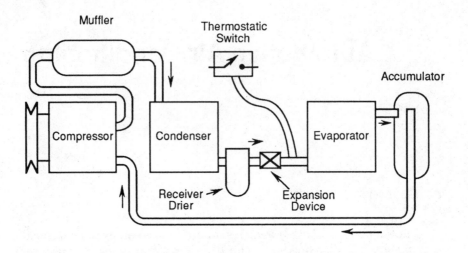

Figure 7.1   Basic Vehicular Air Conditioning System

The high pressure fitting is usually a smaller thread diameter which prevents accidental connection of the charging hose. Adapters which permit a hose connection are available.

Older automotive air conditioning systems were sometimes equipped with shutoff valves which contained the access fittings, similar to those used on some central air conditioning systems. These valves are generally not used on most vehicles in service today.

The Schrader access valve used on most air conditioners contains a spring-loaded stem which seals the valve and prevents leakage of refrigerant, much like a tire valve. The manifold gauge set hoses and high-pressure adapter are equipped with a center pin which depresses the valve stem when the hose or adapter is installed. When the service procedure is completed and the hose removed, valve caps should be installed over the access valves. This will keep the valves clear of dirt and will prevent refrigerant seepage should a valve seal be faulty.

When an adapter is connected to the system, the charging hose and manifold gauge must be attached to it at all times. If not, refrigerant gas, liquid, and oil will shoot out of a charged system with great force since the adapter depresses the valve stem and holds it open.

Perhaps the most obvious difference between a residential line powered air conditioner and one found in an automobile is the compressor, which is not a hermetic unit. Present automotive technology precludes the use of an electrically driven hermetic compressor because of the extremely high electrical power requirement, 2 or 3 kilowatts, that would be needed to run such a system. Automotive alternators cannot supply this level of energy without being far too bulky to fit under a vehicle's hood.

The practical alternative to a hermetic compressor is the belt-driven unit, which derives its power directly from the engine. Since the shaft of the compressor must pass from the inside of the mechanism to the outside pulley, a rotating seal is required to ensure integrity of the refrigerant system.

Another big difference between the automotive and hermetic system is the use of reinforced flexible hoses, which are required to provide the necessary mechanical isolation between the engine-mounted compressor and chassis components. This type of construction precludes the use of R-22 as a refrigerant, which may exhibit excessive leakage and cause deterioration of the hoses. In its place, the less efficient R-12 is used. With the curtailment of R-12 being used as a refrigerant, new products will be available as a replacement for the automotive market. Before the year 2000, all automotive air conditioning systems in production will use new, environmentally safer refrigerants.

Vehicular air conditioning systems must often operate under adverse and varying conditions as compared to AC powered units. The biggest difference is in the speed or rpm of the compressor, which will vary from a very low value at idle to extremely high rates during acceleration and at a highway speed. The automotive engineer must design the system to provide sufficient cooling under all conditions of engine speed.

Varying vehicle speed also affects condenser cooling, which relies on the forward momentum of the vehicle to provide air flow through the cooling fins. Most vehicles manufactured today are front-wheel design with transversely mounted engines. These use electrically operated fans that kick in when normal air flow is not sufficient to provide the necessary condenser or radiator cooling.

Because of the unique conditions under which an automotive air conditioning system is operated, it may often be called into service during relatively cool ambient conditions. It is not unusual for an automotive system to be operated at temperatures far below 70 degrees F (21 degrees C), to provide automatic dehumidification of the air that is directed to the windshield during a defogging and de-icing function. A residential unit operated under such conditions may very well form frost on its evaporator coils, due to the abnormally low operating pressures. In the automotive system, this condition cannot be allowed to exist since frost buildup on evaporator coils prevents heat transfer and dehumidification. This would reduce passenger comfort and be a safety hazard.

7.2 Refrigerant

All automotive air conditioning systems manufactured at the present time employ R-12 as the refrigerant. However, this product, a chlorofluorocarbon (CFC), is now being taxed by the federal government and will be totally phased out of use in the future. Refrigerant manufacturers are presently developing new, environmental-friendly refrigerants which can be used as a replacement for R-12, both in new systems and as a retrofit to older ones. General Motors Corporation, for example, will begin using a replacement refrigerant in some 1993 models, and expects to totally phase out the use of R-12 by the 1996 model year.

R-12 is available in various sizes from the 14 ounce container to a 145 pound (66 kilogram) steel cylinder. This refrigerant is usually packaged in a white color coded container. The air conditioning and refrigeration industry has specified specific container colors for various refrigerants to help personnel and avoid use of the wrong product in a given application. Manufacturers are not required to use the color codes, but most do.

The small R-12 container is convenient to use for topping up a system that is slightly low on refrigerant, or as a readily portable refrigerant supply. When one can of refrigerant is not sufficient to accomplish the desired task, it then becomes an inconvenient (and more costly) method of charging a system.

Fourteen ounce cans of refrigerant require the use of a piercing type valve assembly (Figure 7.2), which allows the charging hose to be connected to the container. The valve is first set to the closed position, then assembled to the can of refrigerant as it pierces the top. A rubber gasket provides a seal so that refrigerant does not escape into the air.

Figure 7.2   R-12 Container with Piercing Valve

Most service technicians will use a larger container of refrigerant, such as a 30 or 50 pound (14 or 23 kilogram) disposable can, or a refillable 145 pound (66 kilogram) cylinder.

The larger sizes are more convenient to use and often offer cost savings over the small containers. These sizes are equipped with hand-operated valves and a 1/4 inch SAE fitting which matches the charging manifold hose connection. The 30 and 50 pound containers may not be legally refilled with refrigerant. This is a DOT safety requirement, and should be observed.

7.3 System Capacities

It may not be apparent, but the heat load presented by a vehicle to its air conditioning system can often match that of a small house or apartment. At the very least, an automotive air conditioning system will have a capacity of 12,000 BTU (1 ton) for a small import, but is often rated 1 1/2 or 2 tons for a full-size vehicle.

Some vehicles, such as stretch limousines and camping vehicles or vans, utilize a second evaporator section which is used to cool the rear section of the vehicle. One compressor is used to supply liquid refrigerant to the front and rear evaporators, each of which contains its own thermal expansion valve. In this way the heat load presented to each section by its blower will control the demand of liquid refrigerant from the compressor/condenser section of the vehicle. Such an air conditioning system can have a total capacity of as much as 4 tons of cooling!

The requirement of such a relatively large cooling capacity in a vehicle is brought about by the severe conditions to which it is subjected. Vehicles are often parked for hours under overhead sun exposure when the outside ambient temperature can easily exceed 90 degrees F (32 degrees C), resulting in an interior temperature of as much as 140 degrees F (60 degrees C). In order for the system to be able to respond to this large heat load in as short a time as a few minutes, an extraordinary amount of cooling power is required. Additionally, because of the large expanse of glass area in most vehicles, the heat load due to sun exposure during midday hours requires at least 1 to 2 tons of cooling capacity.

The capacity of an automotive air conditioning system is a function of the rpm of the compressor, which is driven by the engine of the vehicle. Full system capacity is not available at idle or low cruising speeds less than 30 MPH (19 kilometers/hour). This means that when testing vehicular air conditioning systems for performance, the service technician must set engine rpm to at least 1500 or more.

Under high ambient temperature operating conditions, the power required by an automotive air conditioning compressor can be as high as 2 horsepower for each 12,000 BTU of cooling capacity at highway speeds. For this reason a drive belt, in good condition and correctly tightened, is mandatory to ensure proper operation of the system.

Under idling conditions, the load of the compressor can often be a formidable load for small 4 cylinder engines. Some manufacturers include control components (such as solenoids) which increase engine idling speed during air conditioner operation. At least one import manufacturer designed a system which disconnected clutch power to the compressor during idling, to remove the load on the engine. Computer controlled engines manufactured today automatically set idle rpm correctly as the compressor cycles on and off.

Vehicle manufacturers supply performance data for the air conditioning systems designed into their products. Such data, available in shop manuals, provides a tabulation of evaporator outlet air temperature versus various outside ambient temperature levels. Such measurements will vary over a wide range depending upon the vehicle, engine rpm, and outside temperature and humidity, but will usually be within the range of 40 to 50 degrees F (4.4 to 10 degrees C).

7.4 Types of Systems

Automotive air conditioning systems have been in production for several decades, and have seen several stages of evolution as air conditioning and automotive technology advanced. The original designs, some still being used today, consisted of a cycling clutch compressor which was controlled by a thermal switch mounted near the evaporator to sense its temperature. When the temperature of the coils approached the freezing point, compressor operation was interrupted until the evaporator coils warmed up. This system operated satisfactorily, but during operation on very cool, humid days the temperature control precluded satisfactory dehumidification of passenger compartment air.

The next step in evolution, used by General Motors, was the pressure-controlled evaporator, which used a suction throttling valve or POA (pressure operated absolute) valve to regulate evaporator pressure so that it could never go below 30 PSI. Compressor operation was continuous (with its penalty of lower fuel economy), but no frost could form on the evaporator coils because refrigerant evaporating temperature could not go below the freezing point of water. This design permitted air conditioner operation and dehumidification during any season without the possibility of frost formation on the coils or loss of effective performance.

Some automotive manufacturers, including Chrysler Corporation, used a similar approach to the pressure-controlled evaporator, but designed the pressure control to operate in the 24 to 30 PSI range. These systems, under certain temperature and humidity conditions, could generate a buildup of frost on the evaporator coils if the evaporator pressure was less than 30 PSI and the heat load was not sufficient to prevent formation of ice.

The evaporator pressure control system was used for many years, until the reality of the oil embargo crises changed automotive design philosophy. It was replaced by the next evolution of automotive air conditioning systems, called C.C.O.T. (cycling clutch expansion orifice tube), pioneered by General Motors and still in use today. This system eliminated all moving parts of the sealed refrigeration system, except the compressor. The expansion valve and evaporator pressure valve were eliminated. Operation of the system was controlled by an expansion orifice tube (Figure 7.3), which metered liquid refrigerant into the evaporator at a controlled rate. Evaporator temperature control was provided by either a temperature-operated or pressure-operated switch which controlled the compressor and caused it to cycle on and off in accordance with the cooling demands placed upon the system. This system had three advantages over previous designs: It was many times more reliable due to the elimination of the trouble-prone expansion and evaporator valves, fuel economy was significantly improved because of the cycling clutch, and elimination of two costly valves lowered its manufacturing cost.

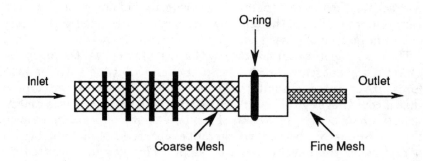

Figure 7.3   Expansion Orifice Tube Used in C.C.O.T. Systems

The refinement of the C.C.O.T. system eliminated the sight glass from the system since diagnosis of system operation was as simple as feeling by hand the temperature of the evaporator inlet pipe and body of the accumulator. This provided virtually instant diagnosis of an automotive air conditioning system.

Although there are many automotive manufacturers worldwide today, all vehicular air conditioning systems are in one form or another similar to those discussed above.

7.5 System Blower

The heater blower in the vehicle is also a part of the air conditioning system. However, when called upon to deliver cooled air instead of heat, its work load is increased significantly. The reason for this is that cool air is denser than warm air, and therefore harder to move. Additionally, the high BTU output rating of the air conditioning system requires a far greater air flow, in the order of 400 cfm per ton of cooling. Because of these factors, more powerful blower motors are required in vehicles that are equipped with air conditioning.

Air conditioning requirements in a vehicle vary over very wide ranges because of possible use during most months of the year when outside temperatures can be cool, as well as hot. In order to allow the vehicular air conditioning system to respond to this varying cooling demand, the blower is usually designed as a three- or four-speed unit. Automatic climate control systems may employ an infinitely variable speed control. This permits minimum cooling capability at low speed, full capacity at high speed, and one or more steps in between. Blower speed control is usually accomplished by a panel-mounted user control switch which cuts in one or more resistances wired in series with the blower motor. Since these resistors dissipate a significant amount of heat when in operation, they are usually placed in the blower air stream for cooling.

The high-capacity blower used in automotive air conditioning systems requires a significant starting and running current when operated at high speed. Blower circuits may be fused at 20 amperes or more, which provides some insight as to their normal operating current level, 10 or more amperes. A well designed blower circuit will include a high speed circuit that uses the contacts of a relay to bypass the heavy starting and running current of the motor. This helps prolong panel switch life. Designs which do not include a high-speed relay circuit are subject to switch failure, which may be evidenced by no blower operation at all, or none at high speed. Figure 7.4 illustrates a blower circuit which was typically used on many General Motors vehicles.

The blower circuit may be designed to operate on low speed at all times (after engine warm-up) unless a higher speed is chosen by the driver. At start-up in cold weather, before the engine has warmed up somewhat, an ambient temperature switch defeats the low blower speed circuit. When high speed is selected, the panel switch operates the high-speed blower relay which is designed to handle the very large inrush current of the blower motor. The high-speed circuit may be protected by its own fuse, located in the engine compartment of the vehicle.

Failure of the blower to operate at any or all speeds may be caused by a defective panel switch, blower relay, fuse, or blower motor. Defective speed control resistors, usually located in the cool air stream of the blower, also may be cause for failure of one or more blower speeds.

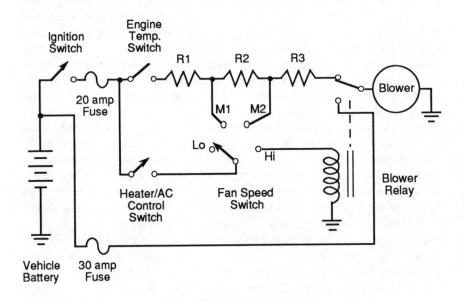

Figure 7.4   Typical Four-speed Blower Motor Circuit
Using High-speed Relay

7.6 Automatic Temperature Control

An automatic temperature control system will perform all of the system adjustments that are required to provide total human comfort during any season of the year. These functions include setting the various air doors to direct the heated or cooled air flow to the floor for heating, or to the dashboard for cooling. Defrost or defogging operation is enabled by a user-controlled panel control switch. The temperature of the air may be modified by a proportional mix of heated and cooled air in accordance with the setting of the thermostat control and the heating/cooling load on the system.

Blower speed is automatically controlled, either by means of an automatic stepping switch which selects the optimum air speed or by a solid state proportional control with infinite resolution.

Manual controls which permit the user to override automatic blower selection speed are located on the control panel, as is a thermostatic control for the desired interior temperature selection. An economy mode control switch allows the user to disable compressor operation (for improved fuel economy) during cooler weather when air conditioning is not required.

Automatic temperature control systems will include an outside air temperature monitoring device which allows the control circuitry to properly anticipate proper selection of system parameters. For example, if both the outside and interior air temperatures are much higher than the thermostat setting, the system blower will be activated immediately upon start-up of the vehicle regardless of engine temperature. Conversely, if the outside and inside air temperatures are below the selected setting, the blower is inhibited from operating until the heater water temperature has risen to a suitable level.

Defrost or defogging is controlled by the user through a panel switch. This causes most of the intake air to be directed to the windshield. During this mode of operation in mild weather, the compressor is operated to provide dehumidification of the air for best defogging results. If the outside air temperature is very low an ambient temperature switch or control circuitry protects the compressor by preventing its operation.

7.7 Vacuum Motors

Because of the multitude of mechanical controls which must be actuated in a combination air conditioning and heating system, vehicle manufacturers have relied on a simple device called a vacuum motor, illustrated in Figure 7.5. This is simply an assembly which contains a diaphragm that is exposed to vacuum, and ambient air pressure, to produce a linear motion that can be used to perform a dedicated task. An example of this is an air door which directs conditioned air either to the floor or panel outlets of a vehicle.

Since the available vacuum produced by gasoline-powered engines is in the range of 10 or 15 inches of mercury, this negative pressure of about 5 PSI (34 KPa) can exert a pull of over 30 pounds on a 3 inch (7.6 cm) diameter diaphragm. Such force is sufficient to actuate a relatively complex mechanism, even if it must overcome a spring return device.

Diesel engines do not produce vacuum as in a gasoline-powered vehicle. For these systems, a separate vacuum supply is provided by the manufacturer of the vehicle.

The use of several vacuum motors in the air conditioning vehicle allows the manufacturer to place the various mechanisms at the optimum location without regard to their proximity to the control panel. A set of vacuum hoses can then be used to deliver the desired vacuum to each of the motors as necessary. Vacuum control is accomplished by a rotary assembly which directs the vacuum supply

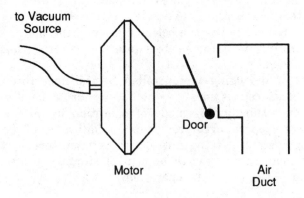

Figure 7.5   Diaphragm Motor Driven by Vacuum Actuates Air Door

to the desired hoses. As indicated in Figure 7.6, an automotive vacuum control system can be quite complex.

When an automotive air conditioning system is checked, it should be operated in all modes. Proper operation of the air conditioning and heating system depends upon all vacuum motors operating smoothly and promptly. Some vacuum control hoses contain restrictors, which are designed to provide a slow, timed transition from one mode of operation to another.

Vacuum problems that are encountered in an automotive air conditioning system are usually caused by cracked or deteriorated vacuum hoses, or one which has been accidentally (or purposely) disconnected. By inspection of all hoses, the fault often can be quickly located. Many vehicles have vacuum hose routing diagrams noted on a decal placed in the engine compartment of the vehicle.

Not to be overlooked is the vacuum storage container, which maintains a relatively constant vacuum supply for the air conditioning system, regardless of the instantaneous vacuum developed by the

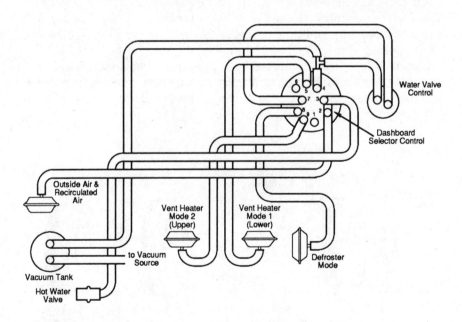

Figure 7.6   Typical Automotive Vacuum Control System

engine. The storage container works in conjunction with a check valve, which prevents stored vacuum from being dissipated into the engine during hill climbing or heavy acceleration.

7.8 Clutch

The compressor is the heart of the automotive air conditioning system. It is belt driven through an electric clutch by the engine of the vehicle, and consumes a significant amount of horsepower. For this reason the drive belt must always be the correct part for air conditioning service, and maintained at proper tension. New belts will lose some tension soon after installation and may require readjustment. Typical belt tension after a belt is broken in will be in the 55 to 75 pound (25 to 34 kilogram) range.

Engine power is transmitted to the compressor by means of a magnetic clutch, which permits the mechanical load of the compressor to be electrically disconnected when air conditioner operation is not required. The clutch assembly (Figure 7.7) is composed of a face plate, hub assembly, bearing, and solenoid coil. Each of these parts is replaceable as a separate item.

Clutch failure may be caused by a defective compressor, which has seized during operation. Should this occur, the drive belt and/or clutch will heat up and eventually be destroyed by the abnormal mechanical

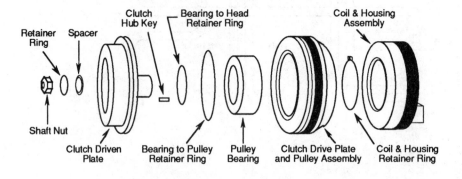

Figure 7.7   Exploded View of Typical Clutch Assembly

stresses placed on these parts. When the service technician detects mechanical damage to either the drive belt or clutch, the possibility of a defective compressor will be indicated. A seized compressor can be checked by attempting to rotate the outer clutch plate by hand when the engine is off and no power is applied to the clutch. If the plate cannot be rotated, the compressor is defective.

Failure of the solenoid coil is rare. It may be checked before disassembly or removal of any parts by measuring its resistance with an ohmmeter. Resistance is measured across the terminals of clutches which are equipped with two male connector pins. Others with a single wire connection are checked by connecting the ohmmeter between the clutch wire (after disconnecting from the vehicle's wiring harness) and chassis ground. Normal readings will be in the 3 to 20 ohm range. An abnormally high value of resistance indicates a defective coil.

If the solenoid coil checks good, but the clutch fails to engage even though it has been ascertained that at least 10 volts power is reaching the coil, the mechanical clearance between the two plates of the clutch may be too great. The manufacturer of the vehicle specifies the correct range of clearance, and this may be adjusted using the proper clutch tools.

Removal of the clutch face plate is accomplished using a special removal tool which has been designed for the particular compressor being serviced. Some compressor face plates may be removed by screwing a large threaded bolt into the threaded hub of the clutch and forcing the face plate out.

Similarly, the pulley and bearing assembly may be removed for service by using tools designed for that purpose. Each part that is removed should be inspected for abnormal wear, and replaced if necessary. Bearings are self-contained assemblies which cannot be lubricated, and must be replaced if defective.

When assembling an overhauled clutch, it is important that the clearance between the engaging plates be within manufacturers' specifications. Excessive clearance may result in a clutch which may not reliably engage; that with too little clearance can cause friction between the parts when disengaged.

7.9 Compressor

The compressor is the hardest working part of the automotive air conditioning system, and is subject to breakdown if not maintained properly. The biggest cause of compressor failure is lack of proper oil charge. A compressor may provide warning of impending failure by noisy operation. If a vehicle exhibits a squealing air conditioner belt or smoking clutch, the compressor has probably seized and must be replaced. A seized compressor can be detected by attempting to rotate by hand the outer clutch plate when the engine is off and clutch de-energized. If the plate cannot be rotated, the compressor is seized.

When a seized compressor is encountered, it should be inspected after removal from the vehicle to determine if it has failed "clean" or "dirty." This is accomplished by draining all of its oil (if any) into a container. If the oil is clean and clear, the system is probably clean. If the oil contains metal particles, the air conditioning system is contaminated and must be cleaned using a purging agent such as R-11. Additionally, the inlet screen to the expansion valve or expansion orifice tube must be removed and replaced or cleaned to ensure that there are no particles blocking the flow of liquid refrigerant. The receiver/drier should also be replaced.

If a compressor has failed clean and contains some oil, make a note of the amount. This information can be used to determine the correct amount of oil charge in the replacement compressor. The vehicle manufacturer specifies the correct amount of oil charge for the system.

The compressor shaft seal may lose its ability to properly contain the refrigerant within the system. Serious shaft leaks, such as indicated by loss of refrigerant charge in less than one season, can be repaired by replacement of the seal using special tools designed for the purpose. Most service technicians will choose to replace the entire compressor assembly rather than attempting seal replacement, especially if the unit is old.

A leaking compressor seal can be detected by a halide or electronic leak detector when the system is charged with refrigerant. This is accomplished by cleaning and blowing away all grease and debris from the front of the compressor, and rotating the outer clutch plate so that a vent opening (if so equipped) is facing downward. After allowing at least 5 minutes resting time, the leak detector can be used to search out a leak at the vent opening.

An oil stain above the compressor on the inside of the hood of the vehicle, or surrounding the clutch assembly, should not be construed as a compressor shaft leak. The seal is designed to seep some oil for lubricating purposes. Replace the shaft seal or compressor assembly only if the detector indicates a shaft leak.

7.10 Compressor Oil

The compressor requires a constant supply of lubrication during its operation. This is provided by the oil charge which travels throughout the system with the refrigerant as the compressor is operated. The components of the system are designed so that relatively little oil will be accumulated throughout and it will always be returned to the compressor along with the refrigerant. Lack of sufficient lubricating oil in the system is the biggest cause of compressor failure.

System oil may be lost, along with the refrigerant, over a period of time through small, minor leaks. Should a system be periodically recharged with refrigerant without replacement of the lost oil, there will come a time when an insufficient supply of oil will be left, resulting in premature wear and possible eventual failure of the compressor.

A catastrophic component failure, such as a blown hose, represents a possible loss of a large quantity of oil. When any component such as a condenser or accumulator is replaced, it may contain a small amount of residual oil which must be added when the new component is installed. The manufacturer of the vehicle provides service data which tabulates the amount of oil which is retained in any component. This quantity of oil must be added to the system to bring the oil charge up to the correct level. If this is not done, the system will have an insufficient oil charge and the compressor will be subject to failure. Typical amounts of oil retained in components are illustrated in Table 7.1.

Some compressors are designed with oil inspection plugs which may be removed (after discharge of all refrigerant) to measure the oil level with a straight piece of stiff wire, much like the dip stick in an engine. If additional oil is required, it may be added through the inspection opening.

Table 7.1 Typical Oil Replacement Chart

Component	Procedure
Condenser	Add 1 ounce (35 cc)
Evaporator	Add 3 ounces (105 cc)
Accumulator	Replace same amount plus 1 ounce (35 cc)
Receiver/drier	Add 1 ounce (35 cc)
Compressor	Add 4 to 6 ounces (consult mfr. specs)

Many compressors are designed without a provision to measure the oil charge while mounted on the vehicle. The only positive way to restore the proper oil charge in such systems is to totally drain the oil from the compressor (and accumulator or receiver/drier if so equipped), and flush the system with a purging agent such as R-11 to remove all traces of residual oil. The system then may be charged to the correct level, using the proper viscosity refrigeration oil, as specified by the manufacturer of the vehicle.

An alternate method to restore oil in a system that has experienced a slow refrigerant leak over a long period of time is to add 2 or 3 ounces of oil. This is not a precise measure, but will provide the necessary lubrication without the danger of a serious overcharge of oil.

7.11 Adding Oil to the System

The easiest way to add oil to a charged air conditioning system is to use readily obtainable pressurized oil containers. These are usually supplied in 4 or 14 ounce cans that contain 2 ounces of oil, with the remainder being refrigerant. This type of oil charge is added to the system using the manifold gauge set in a similar manner as adding refrigerant during a charging procedure, as illustrated in Figure 7.8.

The oil charge is added through the low pressure service connection with the air conditioning system operating. The oil container should be

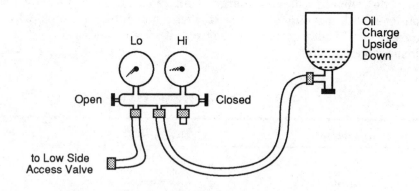

**Figure 7.8   Adding Oil to Charged A/C System
Using a Pressurized Container**

vigorously shaken beforehand to mix the oil and refrigerant, and held upside down during the procedure so that the mixture of liquid oil and refrigerant is forced into the system by the pressure in the container.

Any system that has been opened to the atmosphere may be charged with oil by adding oil directly into an opening, such as the suction port of the compressor or disconnected hose. Rotating the outer clutch plate of the compressor by hand will cause a suction action which will draw in the oil. Oil should not be poured into the suction hose of the system where the compressor will draw in liquid when it is operated. To do so may cause compressor damage. If necessary, turn the compressor clutch plate several turns by hand to distribute the oil.

Oil may be conveniently added to a discharged, sealed air conditioning system prior to the evacuation procedure by taking advantage of the negative pressure produced by the vacuum pump. The setup is shown in Figure 7.9. Refrigerant oil is placed into a clean container and the open end of a charging hose, connected to the high-pressure system access valve, is placed into the container with the oil. The high side manifold gauge valve is closed, and the low side valve opened. As the vacuum pump is operated, the suction developed by the pump will draw the oil into the system.

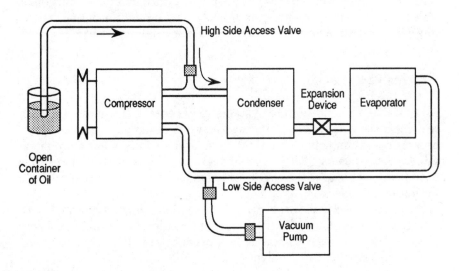

Figure 7.9   Adding Oil to a Discharged System Using the Vacuum Pump

When the oil container is emptied, the hose (and adapter if used) may be disconnected from the high-pressure access valve to seal it. The valve is capped and the vacuum pump allowed to continue operating as it evacuates the system. When total evacuation is completed, the system may be charged in the usual manner.

7.12 Muffler

Many automotive air conditioning systems employ a muffler which is used to reduce characteristic pumping noises from the compressor, and to help minimize sound transfer into the passenger compartment of the vehicle. The muffler is mounted on or near the compressor and is connected to the high-pressure discharge port of the compressor. Some mufflers contain the high-pressure access valve of the air conditioning system.

Mufflers are passive devices and will rarely require any attention from the service technician. The only exception to this is during leak detection, when the muffler, and its associated connections, should be thoroughly checked. The intense vibration produced by some compressors and engines can very well cause metal fatigue of the muffler or its associated parts, resulting in a leak. Any oil accumulation on the muffler assembly may be indicative of a refrigerant leak.

7.13 Ambient Switch

Automotive air conditioning systems may contain an ambient switch which senses outside air temperature and is designed to prevent compressor operation when the outside temperature falls below a predetermined level, such as 40 degrees F (4 degrees C). It is connected in series with the power lead of the clutch, or used in conjunction with a relay, to inhibit clutch operation.

If the compressor is inoperative in a system which is charged with refrigerant, the ambient switch may have failed in an open position. Using a jumper wire to temporarily short out the switch will determine if it is the source of the problem.

7.14 High-pressure and Low-pressure Cutoff Switches

Some air conditioning systems are designed with pressure switches which are designed to prevent compressor operation under abnormal operating conditions, such as might occur when the refrigerant charge is low or there is insufficient cooling of the condenser coils. Operation under low refrigerant charge may starve the compressor of proper lubrication, and cause premature wear and possible failure. An over-heated condenser will result in excessive high-side pressure, and may cause compressor or other component damage.

The low-pressure switch may be located anywhere in the system. A high-pressure cutoff may be located between the discharge port of the compressor and condenser inlet. Some systems employ a pressure cutoff switch located at the receiver/drier. These switches are automatic reset types.

Pressure cutout switches are connected in series with the power feed of the clutch, and prevent compressor operation by opening the circuit. A temporary jumper placed across the terminals of the cutoff switch will determine if it is preventing compressor operation in a system which is properly charged with refrigerant.

7.15 Evaporator Pressure Controls

Automotive air conditioning systems which employ expansion valves to meter the flow of refrigerant may also be equipped with a pressure control to regulate evaporator pressure. Such systems do not use a cycling clutch to prevent evaporator coil freeze-up.

The evaporator pressure control, sometimes called pressure operated absolute (POA) or evaporator pressure regulator (EPR), is located in the suction line or inlet port of the compressor. POA controls are designed to maintain 30 PSI evaporator pressure; EPR valves are rated for pressure control between 24 and 30 PSI.

Proper operation of these controls can be determined by measuring evaporator pressure in a fully charged system during system operation, and ascertaining that the pressure does not fall below the rating of the regulator. For the pressure regulator to operate, the suction pressure of the compressor must be less than the rating of the regulator.

If an access port to measure compressor suction pressure is available, the gauge may be connected to that port to confirm that compressor suction pressure is less than the rating of the regulator. Such pressures are usually about 15 PSI or less when engine rpm is set to 1500 or more and the vehicle blower is running on low speed. If no suction access valve is available, pressure below 30 PSI is indicated if the suction line to the compressor develops a coating of frost (during most temperature/humidity conditions).

7.16 Condenser

The condenser used in automotive air conditioning systems is usually constructed of aluminum tubing, and is located behind the front grille of the vehicle for exposure to the air flow produced by forward vehicle motion. An engine-mounted fan blade, or electrically powered fan motor, provides additional condenser cooling. All the heat energy removed from the passenger compartment must be dissipated by the condenser. This necessitates a clean assembly devoid of leaves, debris, and any other contamination that will impede air flow.

Excessive head pressures in an air conditioning system which seems to be otherwise operating normally may be the result of insufficient air flow through the condenser. The condenser should be inspected for any leaves or other debris that impedes air flow. If the vehicle is equipped with an electrically operated fan, it must be operational. Some vehicle manufacturers stipulate that an auxiliary fan must be used to provide additional air flow through the condenser when checking the system.

Since the condenser is a passive device with no moving parts, the only problem that may be encountered by the service technician is the possibility of refrigerant leaks. Should the manifold pressure gauge, when first connected to the system, read zero pressure, there is a strong possibility that the condenser (or other component) has sprung a serious leak. A thorough visual examination of the condenser and other parts of the system should be performed, looking for the telltale oil stain which will be present in a badly leaking condenser or other part.

If no stain is evident, the system should be partially charged with refrigerant (10 or 20 PSI) and leak tested using a halide torch or electronic detector. Some leaks are relatively small, and greater refrigerant pressure may be required to locate the source of leakage. If no leakage has been detected at the condenser or any other location in the system, a pressure of up to 150 PSI (using carbon dioxide or nitrogen) may be employed to facilitate the leak test. Caution: At no time should a system be pressurized to a level greater than 150 PSI. To do so may result in damage to the system and possible personal injury.

Condensers which have a refrigerant leak can sometimes be repaired. Since the assembly is constructed of aluminum, such repairs are extremely difficult and should be attempted only by those who are experienced with aluminum welding or brazing. A local radiator repair shop may have the expertise and equipment to properly repair an aluminum condenser.

The refrigerant lines which connect the condenser to the rest of the system also are possible sources of leaks, since excessive vibration or an improperly supported condenser assembly and hoses may cause fatigue of the metal pipes. Special attention should be paid to these parts if no leak is found in the condenser itself.

7.17 Condenser Fan

Front wheel drive vehicles have engines which are transversely mounted. This type of design precludes use of the common fan blade which is belt driven by the engine to provide air cooling to the condenser and radiator of the vehicle. In order to properly cool the air conditioning condenser during zero or low vehicle speed on front wheel drive vehicles, an electrically driven motor and fan are employed. The motor operates intermittently in accordance with the cooling requirements of the engine and air conditioning system.

A thermal sensing switch is used to monitor temperature of the condenser, or possibly also the radiator and coolant, and operate the fan motor when required. A relay may be used to carry the relatively heavy load current of the fan motor. Some air conditioning systems will have more sophisticated controls which automatically power the fan motor during compressor operation.

Proper operation of the condenser fan motor will be indicated by the level of head pressure developed by the compressor, which should be within manufacturers' specifications. This pressure should not exceed 300 PSI under any operating condition. Some manufacturers specify that when operating and testing the air conditioning system of a vehicle that is parked, an auxiliary fan should be placed in front of the condenser air inlet to simulate the air flow normally developed by the normal forward motion of the vehicle.

Insufficient condenser cooling can be determined by carefully feeling by hand the temperature of the refrigerant liquid line feeding the evaporator restricting device. Any line which is "red hot" indicates an abnormal situation which must be corrected. Often such a condition will be indicated by a squealing air conditioner drive belt, which cannot handle the excessive load on the compressor.

If the condenser fan motor fails to cut in soon after the compressor has been in continuous operation, the thermal sensing switch may be checked to see if its contacts are closed. The fan motor relay is another possible source of trouble, and is most easily checked by replacement with another part which is known to be good. Finally, the motor itself can be checked by disconnecting its power lead from the vehicle's electrical harness and applying 12 volt power directly from the vehicle battery to the motor. It should run instantaneously.

7.18 Receiver/Drier

An air conditioning system which employs an expansion valve for refrigerant control must also be equipped with a canister, called a receiver (Figure 7.10), to store the liquid refrigerant until it is to be delivered to the evaporator. The expansion valve allows refrigerant to flow to the evaporator only as fast as necessary to maintain evaporator temperature. This necessitates the use of a storage location for any excess liquid refrigerant coming from the condenser.

The receiver is also a convenient component that can be used to hold the desiccant, which is a substance that attracts and holds moisture. The desiccant is placed in a bag which rests at the bottom of the receiver. The combination of these two functions results in the name receiver/drier for this component of the air conditioning system.

Some receiver/driers also are equipped with a sight glass, which allows the service technician to observe the flow of liquid refrigerant as it leaves the container. A steady stream of bubbles during operation of the system indicates that the refrigerant is composed of both liquid and gas, and that the charge in the system is probably insufficient. Other faults, such as a blocked line before the receiver/drier, can also cause a stream of bubbles, but this condition is rare and is easily detected as an abrupt temperature change at the point of restriction.

Many receiver/driers are equipped with a high-pressure relief valve, which will bleed refrigerant safely in the event that system pressure exceeds a safe level. The relief valve will probably never be actuated during the lifetime of the vehicle, but it is included in the system as a safety measure.

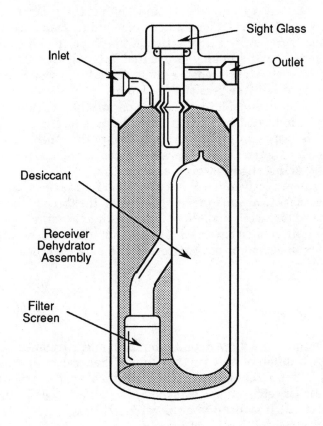

Figure 7.10   Typical Receiver/Drier

Receiver/drier replacement is necessary when it is totally saturated with moisture as might happen if the system is left open to the atmosphere for a sustained amount of time. It must also be replaced if it develops a leak or if it becomes clogged for any reason. The sight glass is sealed with an O-ring and if not properly secured may leak refrigerant. During any leak test of the system, the receiver/drier should be thoroughly checked for any possible leaks.

In an emergency, a water saturated receiver/drier may be returned to service by removing it and baking the component at 120 degrees F (49 degrees C) for an hour or more to evaporate and drive out the moisture. After reconnection of the drier, the air conditioning system must be thoroughly evacuated to at least 29.5 inches of mercury (at sea level) to ensure that virtually all water is removed from the drier and the rest of the system.

A clogged receiver/drier that does not permit the free flow of refrigerant may be indicated by low system pressures and lack of sufficient cooling. Its body may feel abnormally cool to the touch. Normally, a receiver/drier assumes the temperature of the liquid refrigerant flowing through it, and is slightly warm to the touch.

Replacement receiver/driers are available from various suppliers. It is not necessary to use an exact replacement if the threaded connections are identical to the original. Note that the direction of refrigerant flow through the drier must be correct. Most manufacturers use different size fittings for inlet and outlet to prevent the part from being installed backwards.

7.19 Accumulator

An accumulator is a storage container for liquid refrigerant, but has a significantly different function than that of the receiver/drier. Automotive C.C.O.T. air conditioning systems which employ the expansion orifice tube for refrigerant control operate with a "flooded" evaporator. This is in contrast to the "dry" evaporator of the expansion valve control system in which all liquid refrigerant is boiled off before it exits the evaporator.

A flooded evaporator is one in which the supply of liquid refrigerant is greater than that which can be totally vaporized by the time it reaches the evaporator outlet. Using the technique of flooding the evaporator with excess liquid refrigerant ensures maximum cooling capability of the evaporator.

Since there will be some liquid refrigerant discharge at the evaporator outlet, it must be stored somewhere to prevent it from reaching the suction inlet of the compressor, where it can cause serious damage. This is the function of the accumulator, which is located between the evaporator and suction inlet of the compressor. Any liquid refrigerant from the evaporator falls to the bottom of the accumulator and boils off as the gaseous refrigerant is discharged at the top.

As with the receiver/drier, the accumulator is equipped with a bag of desiccant to trap and hold any moisture that is present in the system. Many accumulators are also equipped with the low side access fitting which is used for charging the system with refrigerant.

Accumulators are essentially trouble-free components. The only time that replacement is necessary is when the desiccant is saturated with moisture, or the container develops a refrigerant leak. When the accumulator is replaced, any residual oil that is contained in the old part must be replaced with fresh refrigerant oil in the new one to maintain proper system lubrication. One additional ounce of oil should be added to account for that which saturates the desiccant bag.

Since the accumulator can store a significant amount of system oil, it is necessary to remove and drain such oil whenever it is necessary to measure total system oil charge. This procedure is usually done only when replacing the compressor, to ensure proper oil charge when the system is placed back in operation.

Air conditioning systems which employ the expansion tube orifice and accumulator are extremely easy to diagnose. The procedure is to feel the temperature of the top of the accumulator and the inlet pipe of the evaporator by hand, with the system operating and blower on high speed. If they are both quite cold and essentially equal in temperature, the refrigerant charge is satisfactory. Systems which exhibit a warm accumulator body will almost always be low on refrigerant. If the inlet pipe to the evaporator is warm, the system may be totally out of refrigerant or the expansion orifice tube may be clogged.

7.20 Expansion Valve

Thermal expansion valves used in automotive air conditioning systems perform the same task and operate in the same way as in residential or commercial units. They are thermally operated, employing a sensing tube which is used to monitor evaporator outlet pipe temperature and control a needle valve that meters the proper amount of refrigerant.

Some vehicular air conditioning systems employing expansion valves are designed with provision for automatic compressor clutch cycling to prevent evaporator coil freeze-up. A thermostatic switch, placed on or near the evaporator coil assembly, monitors its temperature and de-energizes the clutch as required. Other systems operate with continuously running compressors that are designed to maintain evaporator temperatures above 32 degrees F (0 degrees C) under most operating conditions.

When the evaporator temperature rises due to increased heat load, the expansion valve opens to permit a greater flow of refrigerant and increased cooling. Should the heat load on the evaporator decrease, the resulting lower core temperature will cause the valve to close. Thus, the valve opening is always at a point of equilibrium, in which the amount of refrigerant flowing into the evaporator is held at a rate that is just sufficient to maintain proper temperature.

Many expansion valves are equipped with an inlet screen which is designed to keep metal particles or dirt of any kind from entering the valve and causing clogging or erratic operation. In a severe situation, the screen could become completely sealed, simulating a valve that is stuck closed. The only cure for this condition is to discharge the system of refrigerant, remove the valve if necessary, and replace or clean the inlet screen.

The sensing element of the expansion valve should be firmly clamped to the outlet pipe of the evaporator, and covered with insulation. The condition of the sensing bulb mounting, and insulation, should be checked on any system which has insufficient cooling or exhibits abnormal pressure levels.

An expansion valve that is stuck in a closed position or which has a clogged inlet screen will starve the evaporator of refrigerant, resulting in a warm evaporator coil. System pressures will be abnormally low. A valve that is stuck open will allow excess refrigerant to enter the evaporator and cause abnormally high system pressures.

Some expansion valves can be checked during system operation by temporarily removing the sensing element from the evaporator outlet pipe and alternately subjecting it to a bath of ice water and heat. The low pressure manifold gauge reading should decrease as the valve closes with the ice water bath, and increase as the sensing element is warmed by hand. Any valve which does not respond to the simulated temperature changes as described is not operational and must be removed from the system for repair or replacement.

7.21 Expansion Orifice Tube

The expansion orifice tube is used in most domestic vehicles on the road. Such a system can be recognized by the large accumulator which is usually located somewhere near the firewall of the vehicle. These systems, called C.C.O.T., employ a cycling clutch which is controlled by either a pressure or thermostatic switch that senses evaporator condition. Other than the compressor, there are no internal moving parts in this system.

The expansion orifice tube is a passive device that meters liquid refrigerant into the evaporator at a controlled rate in accordance with high- and low-side pressures. The only time it will require service is when its inlet or outlet screen becomes clogged due to metal particles or other contamination that may flow through the system. Such a condition is often caused by failure of the compressor. Whenever compressor replacement is required, the expansion orifice tube must be removed from the system, checked for contamination, and replaced if necessary.

A clogged expansion tube orifice will be evident in a fully charged system by lower than normal compressor suction pressure. The evaporator inlet pipe, normally cold during compressor operation, may be warm to the touch since little or no refrigerant is passing through. The accumulator will be warm to the touch, and the compressor may cycle excessively. When the engine is shut off after air conditioner operation for a few minutes, the normal sound of residual liquid refrigerant flowing through the expansion tube will be absent.

The expansion tube orifice is located in the liquid line at the evaporator inlet. To gain access to the tube, the O-ring connection is disassembled, using two open-end wrenches, and the parts separated. The expansion orifice tube is positioned deep within the evaporator inlet pipe and may be removed with a special tool designed for the purpose, or long-nose pliers. The tube may be difficult to remove, and care must be taken not to damage the delicate plastic part.

When reinstalling the tube, be sure to lubricate its O-ring with a small amount of clean refrigeration oil, and insert it with the shorter screen end first. Push it into the tubing until it reaches a firm stop. Examine the connector parts before assembling, and be certain that a new O-ring is in place. A small amount of clean refrigeration oil placed on the ring will help ensure a leak-tight connection. Care must be taken to avoid cross threading the parts during assembly.

7.22 Thermostatic or Pressure Switch

A C.C.O.T. air conditioning system relies on a thermostatic or pressure switch to cycle the compressor in accordance with the heat load on the evaporator. A switch that has failed in an open position will prevent the clutch from being actuated. Failure in the closed position will result in freeze-up of the evaporator core, as evidenced by sustained suction pressures of less than 30 PSI.

The thermostatic switch senses temperature at the evaporator inlet by means of a capillary tube clamped to the pipe. The pressure switch senses refrigerant boiling temperature by measuring evaporator pressure, and controls compressor operation in the same way as a thermostatic switch.

Operation of these switches can be checked by setting the system for normal cooling and low blower operation. The compressor should cycle on and off in 5 minutes or less. If not, ascertain that the switch capillary tube (if so equipped) is clamped securely to the evaporator inlet pipe. Defective switches are not repairable and must be replaced.

7.23 Hot Water Shutoff

Many vehicles employ a hot water shutoff valve which prevents circulation of engine coolant through the heater core when the air conditioning system is set for maximum cooling. The valve is usually vacuum operated, and is actuated if the A/C control panel is set to "maximum," or the temperature control is pushed to the coldest position.

Complaints of insufficient cooling during very hot weather may be the result of a defective hot water shutoff valve, which permits the heater core to remain hot. Possible leakage of heated air, mixing with the cold air conditioned air, will result in a poorly performing air conditioning system.

Shutoff valve operation can sometimes be visually observed as the vehicle air conditioning and heater controls are alternately set for maximum cooling and heating. The vacuum supply to the valve may be checked by using the low-pressure gauge of the manifold, if a suitable adaptor to make the connection is available. If not, a finger placed across the open end of the disconnected line will indicate if there is any vacuum there. Some valves are designed to open with vacuum; others close when subjected to vacuum. They automatically return to their normal position when the vacuum supply is removed.

Valves that are sticky must normally be replaced, but sometimes erratic operation may be caused by old, worn-out antifreeze solution which has lost its lubricating qualities. Flushing the cooling system and adding new antifreeze can sometimes restore proper valve operation.

An improperly operating temperature control bowden cable can also result in poor cooling performance if it fails to completely shut off the flow of hot air from the heater core. One way to check for proper cable adjustment is to briskly move the control lever to each end of its travel. The sound of the air mix door, hitting its stop at each end, should be evident.

7.24 Leak Testing

When an automotive air conditioning system loses part or all of its refrigerant charge over a period of time which is less than one season, it should be checked to determine the source of the leak. Systems which require recharging every few years are not considered to be leaking. The compressor seal is designed to seep some refrigerant and oil for lubrication purposes, and those systems which are used very infrequently (especially over the winter season) will lose more refrigerant over time than those which are in constant use. The owner of the vehicle should be made aware of this fact so as to help keep leakage to a minimum.

An oil stain at or near the compressor clutch is not necessarily an indication that the compressor seal is defective. Only when the leak detector positively indicates a leak should the compressor or seal be replaced.

The most sensitive leak detection instrument is the electronic detector, which can respond to a refrigerant leak as small as 1/2 ounce per year. This detector employs a sensing head which is connected to the instrument by means of a flexible cable or conduit. Any refrigerant entering the sensor causes an audible response or meter deflection if so equipped.

The halide leak detector is not as sensitive as an electronic device, but when properly used can detect leaks as small as a few ounces a year. Best sensitivity is obtained when the flame size is kept as small as possible. The halide detector will respond to a contaminated atmosphere; clear the area of refrigerant fumes, if necessary, so that the flame is essentially colorless before starting the leak-searching procedure.

The halide detector uses a propane flame that receives its air supply through a sniffer tube. The flame heats up a copper plate which is used as a reactor, and the detector is ready for use when the reaction plate glows red. If the air supply tube feeding the propane burner is passed over a leak so that it contains a small amount of refrigerant, the color of the flame changes from a pale blue to yellowish or bright green, depending upon the concentration of refrigerant. A large leak will cause a bright blue flame, and that produces a very toxic gas (phosgene). If this happens, the leak detector should be immediately removed from the source of refrigerant. It is best if the halide leak detector is always used in a well ventilated area so that any fumes produced by the torch are quickly dissipated.

Although various types of refrigerant oil dyes are available as leak detection aids, it is not recommended that these be used. Their usefulness is questionable, and they may cause a reaction with the oil or refrigerant in the system which results in harmful by-products. Vehicle manufacturers frown upon use of such products, and the owner's warranty may be voided.

All tubing connections and other possible leak points should be checked with the detector. The refrigerant is heavier than air, so approaching the area being tested from below will be helpful. Move the sensing end of the detector slowly, perhaps one inch per second.

Evaporator leaks are difficult to find if they are extremely small. This part is located inside a closed air chamber or plenum and is not readily accessible. Sometimes it is possible to obtain access to the evaporator coil by removing a mounting plate that contains the speed reducing blower resistors, which are placed in the air stream for cooling. When this plate is removed, the leak detector tube or cable may be inserted into the opening to search the compartment for any possible refrigerant leak.

Large evaporator leaks can sometimes be detected from the passenger section of the vehicle. The air conditioning system is first pressurized with refrigerant and the leak detector sensor placed at the air openings in the dashboard. Sometimes setting the blower to low speed helps. Refrigerant that is leaking out of the evaporator coil will find its way into the air outlets where it may be detected.

Very small refrigerant leaks may be difficult to locate in an automotive air conditioning system, especially if one exists somewhere in the high-pressure side. This will produce greater leakage during system operation than when the vehicle is idle. Pressurizing a system to 150 PSI with dry nitrogen can sometimes help in finding leaks. The system cannot withstand pressures greater than 150 PSI, and it must never be pressurized above this level.

7.25 Leak Repair

The automotive air conditioning system is subject to several different kinds of leaks, much different from those of the hermetic system. Many leaks will be repairable, but not all. The technique that is used by the service technician will depend upon the location of the leak and components involved.

The compressor seal is possibly one of the more common points of leakage, since this is a mechanical seal which is subject to wear. As the compressor sees use, it may tend to leak at a greater rate with time. Compressor seal replacement kits are available from the manufacturer of the vehicle as well as the aftermarket. Replacement of this part is not extremely difficult, but special tools are often needed to do the job. The compressor will usually have to be removed from the vehicle for seal replacement. Many service technicians will replace the entire compressor rather than attempt seal replacement, especially if the compressor is original equipment in a vehicle which has seen many miles of service.

A common source of refrigerant leaks is the various rubber hoses which are part of every vehicular air conditioning system. Many times leaks are caused by abrasion of the hose against a metal part of the vehicle. Additionally, when the parts are several years old, the hoses may leak at the points where the manufacturer has crimped a connector fitting to the hose.

Exact hose replacement assemblies may not be available from the manufacturer of the vehicle, especially for older models. However, the aftermarket provides an extensive selection of fittings which can be used to make up any replacement requirement.

In many cases it is not necessary to replace an entire hose assembly. All that needs to be done is to cut the hose at a location which allows the defective part to be discarded, and replace only that portion. Care should be taken to avoid any particles or debris from entering the open ends of the hose. Various sizes of splices are available which enable the service technician to connect a new piece of hose to the cut end of the old one.

Only the correct size barbed hose fitting and clamp should ever be used on automotive air conditioning systems. Any attempt to use a smooth pipe and clamp will result in a burst connection from the high operating pressures, and will be a danger to anyone standing nearby. A typical hose splice with its clamps is illustrated in Figure 7.11.

When a hose develops a leak due to abrasion, it is a simple matter to repair the leak by cutting the hose apart exactly at the location of the leak, and using a hose splice to connect the parts together. When

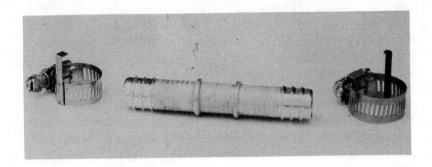

Figure 7.11   Hose Splice

assembling the hose fittings, be sure to lubricate the inside of the hose with clean refrigerant oil before inserting the barbed pipe. If not, insertion may be difficult. The clamps supplied with the splice are designed with a locating tab which is placed against the end of the hose to ensure proper placement.

Leaks that occur at O-ring connectors may be caused by a defective or missing O-ring. Such connectors are very reliable, and they do not need to be over-tightened to properly seal the connection. If a leak is found and the connector parts seem tight, the only way to correct the problem is to discharge the system and disconnect the parts for inspection. Upon reassembly, a new O-ring, slightly oiled with clean refrigerant oil, should be used. Care should be taken not to cross thread the fittings, which can be easily damaged. A small amount of oil placed on the threads may help in assembly. O-ring connectors should be assembled with a backup wrench to prevent twisting of the parts, and tightened to the proper torque as illustrated in Table 7.2.

Leaks that occur in the evaporator or condenser may not be repairable since these parts are constructed of aluminum. However, an experienced technician who is familiar with aluminum brazing or heliarc welding may be able to make a satisfactory repair if the leak is accessible. There are also available epoxy compounds which are designed for evaporator and condenser leak repair. Whenever any repairs are made, the part should be pressurized to about 150 PSI

Table 7.2   Torque Specifications for O-ring Fittings

Aluminum Tube OD	FT-Pounds Torque
1/4 inch 0.64 cm	6
3/8 inch 0.95 cm	12
1/2 inch 1.3 cm	12
5/8 inch 1.6 cm	20
3/4 inch 1.9 cm	25

(dry compressed air is satisfactory) and checked to see that the repair can sustain the pressure.

Some aluminum evaporators and condensers can develop porous areas which contain many minute leaks over a large area. Such components are not repairable and must be discarded.

7.26 Purging

Purging an air conditioning system (Figure 7.12) consists of flushing out contamination of one component or the entire system with an inert gas or refrigerant. Dry nitrogen and R-11 are commonly used for this purpose. Note that R-11 is applied to the system in a liquid state; it should be recovered and recycled. Dry nitrogen has an advantage in that it is economical to use and environmentally safe.

R-11 is available in small containers which are pressurized with nitrogen. An alternative method is to use a suitable refrigerant container which can partially be filled with liquid R-11 and pressurized to about 50 PSI with dry nitrogen gas. The purging agent may be applied to an idle air conditioning system through an access valve, using the manifold gauge set in a similar manner as when charging. The liquid and gas may be allowed to exit and be collected at the other access valve, or at any other point in the system which has been disconnected.

Dry nitrogen has the ability to sweep out moisture that may be in the system, and has an advantage over R-11 in that it does not cause a temperature drop which will thicken the refrigerant oil in the system.

After the purging operation is completed, the system should be reconnected if necessary and evacuated. Any refrigerant oil that was removed during the purging process should be replaced.

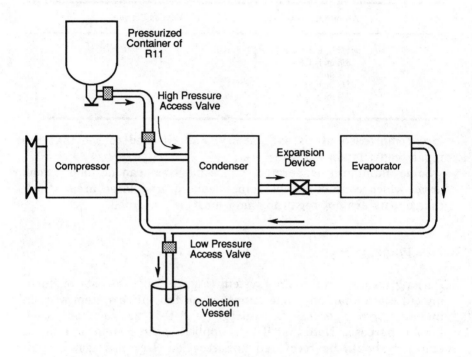

Figure 7.12   Purging a System with Pressurized R-11

7.27 Evacuation and Charging

Caution: Whenever dealing with refrigerant under pressure, safety goggles should be worn. Liquid R-12 refrigerant escaping into the atmosphere will be at a temperature of –21.7 degrees F (–29.8 degrees C) and can cause instant frostbite. Care should be taken to avoid any exposure to liquid refrigerant, especially the eyes. If so exposed, a few drops of sterile mineral oil will absorb the refrigerant, and the eyes should be washed with a weak solution of boric acid. Seek immediate medical attention.

Never remove a charging hose from an adapter connected to an access port of any system that is under pressure. Always remove the adapter from the air conditioning system with the hose and manifold gauge attached. Failure to observe this precaution will result in total loss of all system refrigerant, and may cause personal injury.

Any air conditioning system which has been opened to the atmosphere for any reason must be evacuated of all air, moisture, and other contaminants that have entered the system. Such evacuation should be taken to at least 29.5 inches of mercury (100 KPa) at sea level. If the vacuum pump is not capable of this level, the evacuation may be performed 2 or 3 times after filling the system with refrigerant, to zero pressure level, after each evacuation.

The maximum attainable level of vacuum is dependent upon the elevation or altitude, and is about 1 inch of mercury less than 29.92 inches for each 1000 feet (305 meters) of altitude. For example, at 1000 feet of altitude the maximum level of vacuum will be about (29.92 − 1), or 28.92 inches of mercury (98 KPa).

There may be some confusion in the mind of the service technician when making a decision to totally evacuate a system that has not been opened to the atmosphere, or simply to top it off with the required amount of refrigerant. Evacuation of a system which is low on refrigerant and contains no moisture or contaminants is not only costly, but any R-12 discharged into the atmosphere creates environmental problems. If a refrigerant recovery system is available, the evacuated refrigerant can be reclaimed for use. If refrigerant recovery is not possible, the system should usually not be evacuated before recharging.

If possible, determine the service record of the system. The owner may be the best source of such information. A relatively new vehicle that has never seen air conditioner service but is not seriously low on refrigerant will probably not require evacuation.

Moisture can enter an air conditioning system only by two methods: by exposure to the atmosphere when it is disassembled or opened, or through a leak which is located at any point in the system that has less than atmospheric pressure. It becomes obvious, then, if the factory-installed air conditioning system has always had a reasonable amount of charge, and the low-pressure side never approaches zero pressure under any operating circumstances, then moisture from the outside cannot seep into the system.

Most vehicles use the C.C.O.T. system, which never operates near zero suction pressure because the thermostatic and pressure controls prevent compressor operation when such pressure goes below

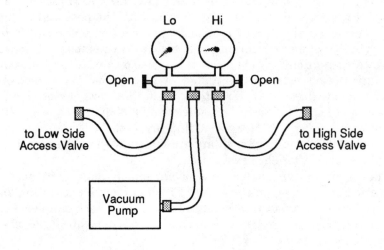

Figure 7.13   Evacuating Both Sides of A/C System

30 PSI. It is extremely unlikely that such systems will have moisture within if they have never been opened to the atmosphere and are at least partially charged with refrigerant. Other systems which employ expansion valves for refrigerant control may at certain times have negative compressor suction pressure. These systems can draw in moisture if there is any air seepage on the low side of the system.

The service technician has the option of evacuating a system (Figure 7.13) using both low- and high-pressure access valves, or using just the low pressure connection. The latter method, illustrated in Figure 7.14, allows the refrigerant storage container to be connected to the setup prior to the evacuation procedure. The charging process can begin without disconnecting any hoses, thus avoiding the necessity of purging such hoses of air.

Many vehicular air conditioning systems use a small size access fitting (3/8-24 thread) for the high-pressure side to avoid accidental misconnections to the manifold pressure gauge. To make a hose connection to this fitting, an adapter will be required. The adapter contains a depressor pin which opens the valve when connected.

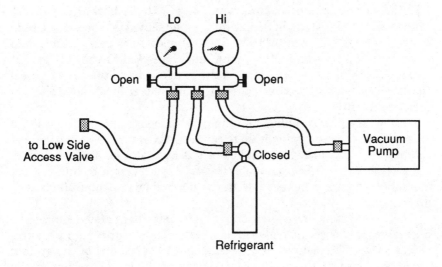

Figure 7.14   Evacuation from Low Side Access Valve

When performing the evacuation procedure, the low-pressure and high-pressure manifold valves should be opened, and the refrigerant container valve (if connected) should be closed. The vacuum pump is operated and the low-pressure gauge monitored to observe the process. Early in the procedure, temporarily close the manifold valves to determine if the system is vacuum tight. If so, open the valves and continue the process.

Evacuation may take anywhere from a few minutes to 1/2 hour or more, depending upon the cfm capacity of the pump and the amount of moisture in the system. If there is substantial moisture, it will have to be totally vaporized (boiled away) by the low-pressure condition before the pump will be able to reach its final level of vacuum. If the level of vacuum in the system seems to reach a plateau of less than maximum pump capability, moisture is being removed. Continue the evacuation process until maximum vacuum capability is attained.

When the system has its final vacuum level, both valves of the manifold gauge set are closed to observe if the air conditioning system can hold vacuum for at least 5 minutes. If it does, it is ready for charging.

If not the source of the leak must be determined and repaired, and the evacuation procedure repeated.

If the evacuation procedure was done with both charging hoses connected to the system, both manifold gauge set valves should be closed to preserve system vacuum. The hose connected to the vacuum pump may then be removed at the pump and connected to the refrigerant storage container. The container valve is opened and the hose connection at the other end, at the gauge set, is momentarily loosened slightly to purge the line of air.

The low-pressure manifold valve may be opened to allow refrigerant to enter the system and bring it up to container pressure. The refrigerant container should always be held upright so that gas, and not liquid, enters the system. (Special adapters which permit liquid charging are available and may be used if desired. Follow the manufacturer's directions.)

When the system has been brought up to container pressure, close the high-pressure manifold valve if it is connected to the high-pressure access valve. This is important—the high-pressure side of the system must be closed to the refrigerant container during operation of the compressor. If not, the container may burst.

With the low-pressure manifold valve and refrigerant container valve open, operate the air conditioning system while observing the pressure gauges. Note that the low-pressure gauge reading drops as the compressor draws refrigerant into the system. The compressor may cycle on and off during the charging process. If desired, an adapter cable may be used to power the clutch directly from the battery to prevent cycling and facilitate the charging process. An alternate method is to temporarily short out the control switch.

If the air conditioning system is equipped with a sight glass, it may be monitored to determine when the charging sequence is nearing completion. This is indicated when the flow of bubbles ceases. At this point another 1/2 pound, not more, of refrigerant may be added to complete the charge.

C.C.O.T. systems may be monitored by operating the blower at maximum speed and checking the temperature of the top of the accumulator by hand as the refrigerant enters the system. When the temperature is biting cold, allow an additional 1/2 pound of refrigerant to enter the system to complete the charge.

A more exact method to properly charge an automotive air conditioning system that has been evacuated is to use a scale to measure the amount of refrigerant, in pounds or kilograms, that is used. The vehicle is usually supplied with a decal which specifies the proper charge of refrigerant. This method of charging is sometimes required on systems that employ expansion valves and are not equipped with a sight glass.

Charging stations which contain scales are available, but any accurate scale may be used. The container of refrigerant is weighed prior and during charging, and the procedure is stopped when the required amount has been delivered to the system.

When charging an automotive air conditioning system, the temperature of the refrigerant container will fall as liquid refrigerant boils off and enters the air conditioning system. This will result in reduced container pressure and lengthen the time required for charging. The procedure may be facilitated by placing the refrigerant in a container of warm water, not over 125 degrees F (52 degrees C), to maintain container pressure. Caution: Never use an open flame to heat a can of refrigerant, and under no circumstances should it ever be subjected to a temperature greater than 125 degrees F. To do so may cause the container to burst.

Caution: If the high-pressure gauge hose has been connected to the high-pressure access port during charging or system analysis, this hose will contain liquid refrigerant which will spew out when the hose is removed. This condition may be avoided through the use of a check valve adapter, which is placed between the high-pressure access fitting and hose. Another method to prevent the loss of liquid refrigerant before disconnecting the hose is to shut down the system and open both low- and high-pressure gauge valves. This will permit the liquid refrigerant in the hose to vaporize safely into the low-pressure side of the system.

7.28 Performance Check

In the final analysis, the only way to determine if an automotive air conditioning system is properly operating is through a performance check. This is a test of the system, under certain operating conditions, in which the system pressures and passenger air outlet temperatures are measured and compared to the vehicle manufacturer's specifications.

Service manuals, provided by both vehicle manufacturers and independent sources, will provide service data for any given vehicle model. It is best to refer to data supplied for the exact model under

Table 7.3 Typical Performance Check

Outside Ambient Temp	70 F 21 C	80 F 27 C	90 F 32 C	100 F 38 C
Comp head pressure PSI	150-190	180-220	240-280	300-340
Evaporator pressure PSI	28-31	28-31	28-31	28-31
Air outlet temperature	36-40 F 3 C	36-40 F 3 C	38-42 F 4 C	41-45 F 6 C

test, but if not available, typical average operating conditions as illustrated in Table 7.3 can be used in place of the manufacturer's performance data.

The performance check should be performed on a fully charged air conditioning system at ambient temperatures of at least 70 degrees F (21 degrees C) or more. The interior of the vehicle should not be much over 90 degrees F (32 degrees C) prior to the start of the test, or the system should be operated until the interior is cool. The vehicle should be in the shade, doors and windows closed, and the air conditioning system set for maximum cooling with high-speed blower operation.

Engine rpm should be set to about 2000 rpm. Some vehicles will require auxiliary air cooling, such as that supplied by a portable fan placed near the front grille of the vehicle, to blow air into the condenser assembly.

Passenger compartment air outlet temperature is measured with an ordinary but accurate thermometer. The readings obtained will be somewhat affected by atmospheric relative humidity level. On very humid days temperature readings above average will be normal; very dry weather will result in lower readings.

The air conditioning system should be operated at least 10 minutes, or until the system and temperature readings have stabilized.

If the passenger compartment outlet air temperature is within the range indicated in the performance data chart but the air conditioning system does not properly cool the vehicle on very warm days, there may be a problem with reduced air flow through the evaporator. An experienced service technician will be able to judge if such air flow is below that of other similar vehicles. Reduced air flow may be caused by dirt or debris clogging the evaporator or a blower motor which is not operating at high speed. If the motor is operating normally, the only way to restore proper air flow is to determine the cause and clean the evaporator coil assembly if necessary.

7.29 Analysis and Trouble Shooting

Analysis of faults in a vehicular air conditioning system is generally not a difficult task. There are certain parameters which should be observed, and these will generally point directly to the section of the unit which is at fault. Many times the problem may be accurately diagnosed without even connecting the manifold pressure gauge to the system to measure pressures. Using sight, sound, and touch, an enormous amount of information concerning the condition of an air conditioning system can be assimilated.

Whenever there is any doubt if an air conditioning system is faulty, the system performance check should be made. The temperature and volume of passenger compartment outlet air, and the operating pressures of the system, will positively indicate a fault when one exists. Many times the system may seem at fault in extremely hot or humid weather, but is not. Some vehicles, especially small, low-priced models, may have marginal air conditioning systems which cannot provide sufficient cooling under extreme conditions.

To analyze a vehicular air conditioning system, it should be initially set for high blower speed and maximum cooling mode. Use extreme care keeping hands, clothing, etc., away from the fan blade and other moving parts of the vehicle.

The following operating conditions should be noted:

1. Blower is operating at high speed. There is a large air flow emanating from all dashboard outlets, and little or none at the floor or windshield. All blower speeds are operational.

2. The compressor is operating in continuous or cycling mode. If cycling, the duty cycle must be sufficient to provide adequate cooling. There should be no unusual compressor noises, and the drive belt should not slip or squeak.

3. The condenser fan motor (if so equipped) should cut in after a few minutes of operation when the condenser heats up, and should ensure a properly cooled condenser.

4. The sight glass (if so equipped) should be essentially clear. A few bubbles now and then do not necessarily indicate a low charge.

5. The evaporator inlet pipe, if accessible, should be felt by hand and should be cold.

6. The top of the accumulator (if so equipped) should be biting cold. The suction line to the compressor should be cool to cold.

7. The discharge pipe of the compressor should be mildly to extremely hot (use care when touching).

8. The liquid line between the condenser and evaporator should be warm, not extremely hot.

The following is a discussion of the various tests and possible symptoms that may be revealed by the above checks:

1. An inoperative blower may be caused by a blown fuse, defective panel switch, blower relay, or blower motor. If one or more speeds is not operational, the speed control resistor assembly should be checked. Lack of air flow through the panel outlets may be caused by an evaporator that is clogged with leaves or other debris, or is icing up due to abnormally low evaporator pressure. Air flow that does not emanate from the correct air outlets in the passenger compartment may be caused by a problem in the vacuum control system. Ductwork should be checked for proper position and fit.

2. An inoperative compressor may be caused by any of the switches which are designed to control compressor operation and cycling. These include pressure switches, and ambient or evaporator thermal switches. The control panel A/C switch, relay, or fuse may also be at fault. Lastly, the compressor electrical connector or wires may be broken or disconnected.

3. The condenser fan motor must operate when the heat developed in the condenser coils is too great to be cooled by natural air flow. This

motor is controlled by a thermostatic switch and relay.

4. A low refrigerant charge will result in a steady stream of bubbles of gas in the sight glass. A system that is fully charged, or one which has little or no refrigerant, will exhibit a clear sight glass.

5. The evaporator inlet pipe is normally flooded with liquid refrigerant which is boiling at the evaporator pressure. As a result, the inlet pipe should be cold. A warm pipe indicates very low refrigerant charge, or a blockage in the system preventing the flow of refrigerant.

6. C.C.O.T. systems operate with a flooded evaporator, and the excess liquid refrigerant is discharged into the accumulator where it absorbs heat as it boils off. The cold gaseous refrigerant may also cool the suction line to the compressor. A warm accumulator indicates insufficient refrigerant charge, which in turn causes too rapid cycling of the compressor.

7. The high-pressure gas discharged by the compressor is heated by compression. A cool discharge pipe indicates very low or no refrigerant in the system, or a faulty compressor.

8. A hot liquid line may indicate insufficient condenser cooling. Check for clogged air passages and condenser fan operation. Some vehicles require auxiliary cooling when operating the air conditioning systems while the vehicle is parked.

By using the information provided by the above checks, it is possible to make a reasonable estimate of the condition of the air conditioning system, and possibly determine the source of a fault if one exists. Since most air conditioning problems are caused by a low refrigerant charge, many of the tests will indicate this as the problem. Connecting the manifold gauge set to the system will then help verify the fault.

Dehumidifiers

8.1 General Information

Dehumidifiers (Figure 8.1) are a specialized form of refrigeration equipment that are designed to condense and collect water vapor that is present in the surrounding air. They do not require ventilation to the outside as do air conditioners, since no heat needs to be removed from the area in which a dehumidifier is operated. The BTU energy developed by the system is used to recover latent heat from the moisture in the air, converting it to water.

Dehumidifiers are used instead of air conditioners for removing moisture from the air for two important reasons. They are usually significantly less costly to operate, and do not cause any appreciable temperature change as do air conditioners. They find extensive use in damp basements, for example, where the temperature is normally cool. The use of a dehumidifier will help prevent the formation of mildew, which can destroy almost any type of soft goods.

In order to understand how dehumidifiers work, the concept of "dew point" should be understood. Dew point is defined as the temperature below which the moisture in the air will condense, and it is equal to the point at which the relative humidity becomes 100%. On a psychrometric chart, this is the wet bulb or saturation temperature line. The amount of water vapor that can be contained in a given volume of air, at a constant pressure, is a function of its temperature. Warmer air can hold more moisture than cold air.

An ordinary glass of ice water will collect moisture on the outside because the air that is in contact with the glass is reduced in temperature below its dew point. Since relative humidity cannot exceed 100%, water vapor is condensed to liquid. The dehumidifier works the same way, except an evaporator coil, cooled by the evaporation of liquid refrigerant, is used instead of a glass of ice water.

The evaporator and condenser in a dehumidifier perform the same task as in an air conditioning unit, where the gaseous refrigerant gives up heat to the air and condenses to liquid in the condenser, and absorbs heat as it changes back to a gas in the evaporator. In order to operate the

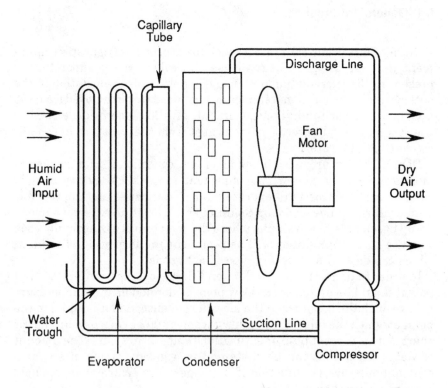

Figure 8.1   Dehumidifier Photo

system with essentially no change in air temperature, the ventilating fan draws air first through the cold evaporator section where it gives up some of its moisture in the form of water and some heat. The drier, slightly cooled air then passes through the condenser coils to condense the high-pressure refrigerant gas as it picks up heat. In this way the flow of air through the unit undergoes essentially no change in temperature.

The electrical energy dissipated in the compressor and fan motors will cause a minor amount of heat to be generated, just as any appliance converts its electrical input to heat. For this reason a dehumidifier which operates in a small enclosed area may cause a noticeable rise in air temperature.

8.2 System Operation

Dehumidifiers are more closely related to refrigerators than to air conditioners since they employ the same refrigerant as most refrigerators, R-12, and operate with similar-sized small hermetic compressors and capillary tubes. The main difference, however, is refrigerant suction pressure and temperature of the evaporator. An optimum design, for example, would consist of a unit with an evaporator temperature that is close to freezing, but not below. This requirement is to preclude any possibility of the formation of frost on the cooling coils. Should this occur, the coils would become insulated, negating the dehumidification process. In order to avoid frost buildup on the evaporator coils, dehumidifiers usually operate with coils that are cooled to the 40 degrees F (4.4 degrees C) range.

Since the suction pressure of a capillary-tube controlled refrigeration system depends to a great extent on the ambient temperature of the air flow across the condenser coils, dehumidifiers must be designed for a nominal evaporator operating temperature that is somewhat above the freezing level when operated at normal room temperatures, or slightly below. This is to ensure that frost will not form on the evaporator coils if the unit is used in an area where temperatures of perhaps less than 60 degrees F (15.6 degrees C) exist. The manufacturer of the dehumidifier will specify the minimum temperature at which the unit will perform satisfactorily.

A psychrometric chart, illustrated in Figure 8.2, may be used to determine the relative humidity level in the air. This chart illustrates the difference in temperature readings of a dry bulb and wet bulb thermometer for various levels of relative humidity. A psychrometer is

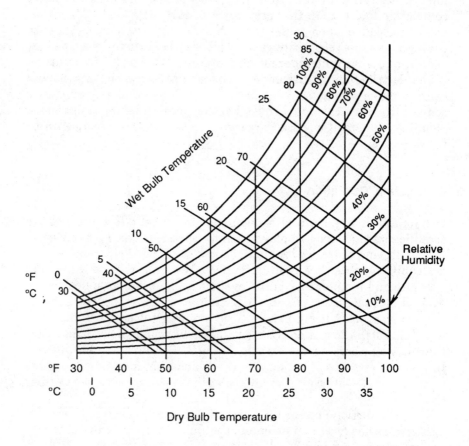

Figure 8.2   Psychrometric Chart

an instrument which contains two thermometers, one with a moistened bulb, that are used in conjunction with the chart to determine relative humidity level. The wet bulb thermometer will always read less than the dry bulb thermometer (except at 100% relative humidity), and the point at which the readings intersect on the chart indicates the humidity.

This chart is used by noting the point where the dry bulb temperature, indicated by vertical lines, intersects with the wet bulb temperature, indicated by the oblique lines. This point will lie at the level of relative humidity indicated in the chart by the curved, constant humidity lines.

A typical dehumidifier may have an evaporator temperature of about 45 degrees F (7 degrees C) during normal operation. By consulting the psychrometric chart for this saturation temperature (wet bulb scale), it can be seen that the relative humidity at a room temperature of 60 degrees (dry bulb temperature) is about 30%. This means that the dehumidifier cannot remove moisture at relative humidity levels that approach 30% or less. This is an important point; a perfectly operating dehumidifier, operating in a low-humidity environment, will remove no moisture from the air because the temperature of the evaporator coils is above the dew point.

Dehumidifiers are rated by the number of pints of water that will be removed in 24 hours under standard conditions, which is an ambient temperature of 80 degrees F (26.7 degrees C) at a relative humidity of 60%. Dehumidifiers are not usually rated in BTU capacity, but they use about 1000 BTU of heat energy (latent heat of vaporization) to condense sufficient water vapor and produce 1 pint of liquid.

8.3 Compressor

Compressors used for dehumidifier service are hermetic types that are generally rated between 1/12 and 1/4 HP, as illustrated in Table 8.1. Such units may draw about half the line current of a small (5000 BTU) window air conditioner. Since the evaporator pressure of a dehumidifier is over 30 PSI, a high back pressure compressor is required.

**Table 8.1   Typical List of Hermetic Compressors
Used in Dehumidifier Service**

Nominal HP Rating	BTU Capacity	Full Load Current
1/12	1131	2.18A @ 115V
1/10	1390	2.30A @ 115V
1/8	1628	2.80 @ 115V
1/6	2084	3.55 @ 115V
1/5	2541	4.22 @ 115V
1/4	2958	5.15 @ 115V

Many dehumidifier units employ hermetic compressors which do not utilize a starting or running capacitor, as those found in air conditioning units. These very low capacity compressors employ a starting relay which is current operated and momentarily connects the start winding of the compressor to the line. Figure 8.3 illustrates a typical current relay starting circuit.

The starting relay contains an armature which is actuated by the heavy inrush current of the run winding of the compressor when power is first applied. This connects the starting winding to the power line and causes a torque to be developed in the compressor motor. When the compressor reaches speed, the running winding current quickly drops to its normal running level. This causes the relay to disengage, opening its contacts and disconnecting the starting winding from the circuit.

Some starting relays are connected directly to the terminals of the compressor. They are secured by pushing the relay assembly onto the compressor connections. Relays of this type are position sensitive since the force of gravity is used to keep the contacts open when the relay is not energized.

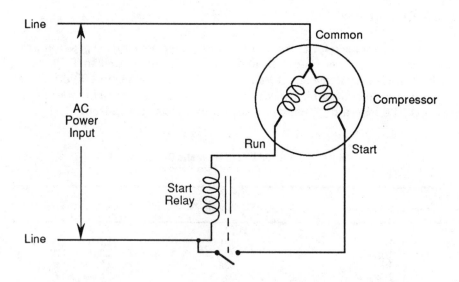

Figure 8.3   Typical Small Hermetic Compressor Starting
Circuit Using Current Operated Relay

Some dehumidifier circuits employ a thermally operated relay which performs the same task as a current relay, but is actuated by heat developed in a thermal element. Other types of circuit designs may include potential relays or solid state circuits.

The compressor must also be equipped with an overload cutout, which provides protection in the event that the compressor does not start when power is applied. The locked rotor current of a compressor is about 10 times its running current, and can quickly burn the windings out if sustained for more than a few seconds. The overload may be part of the starting relay assembly, or may be a discrete component similar to that used on an air conditioning hermetic compressor.

Very large capacity dehumidifiers will have BTU capacities that are similar to small window air conditioners, and may employ PSC compressor motors that require a running capacitor.

8.4 Condenser

The condenser assembly may be manufactured out of steel, copper, or aluminum tubing. It may be painted black to enhance its heat emissivity, and is mounted close to the evaporator.

The high-temperature, high-pressure discharge gas from the compressor is fed into the condenser so that air flow across the coils will cause the refrigerant to give up part of its heat and condense into a liquid. The refrigerant is discharged at the bottom of the coil assembly where it is fed into a capillary tube.

The condenser assembly of a normally working dehumidifier will operate at a warm temperature, regardless of the ambient relative humidity level or dehumidification process. Any unit which does not exhibit a temperature rise in the condenser is most likely low on refrigerant charge or has a restricted capillary tube.

8.5 Capillary Tube

Most dehumidifiers employ a capillary tube as the mechanism by which the low-pressure (evaporator) and high-pressure (condenser) sections of the refrigeration system are separated. It is a low-cost, fixed-restriction device which has no moving parts and provides the necessary metering of the liquid refrigerant into the evaporator.

Failure of the capillary tube, due to restriction, is rare unless it has been subject to mechanical damage. Restriction of the capillary tube may also be caused by moisture in the sealed system, which can form ice and prevent the flow of refrigerant. The length and inside diameter of the capillary tube determine the total resistance to the flow of liquid refrigerant. For this reason, if replacement should ever be necessary, the new part should have the exact same ID and length as the old one.

Many dehumidifiers are assembled with the capillary tube either soldered along the evaporator return tubing to the suction inlet of the compressor, or tightly wound around it. The purpose of this type of construction is to utilize the heat absorbing capability of the superheated refrigerant gas in the suction line, which generally is at a relatively cool temperature. Any heat removed from the capillary tube in this manner will help prevent the phenomenon of flash gas in the capillary tube, in which the liquid refrigerant vaporizes instantly as it leaves the tube. The heat transfer from the capillary tube to the suction line also helps in vaporizing any possible liquid refrigerant in the suction line before it reaches the compressor.

8.6 Evaporator

The evaporator section of the dehumidifier is that portion in which the liquid refrigerant boils as it draws heat from the coils, and becomes cold. The temperature of the evaporator is held below the dew point at high humidity levels, and any humid air which strikes the coils will give up part of its moisture in the form of water.

The evaporator coils are placed in a vertical plane to facilitate the flow of water to the collection tray below, where it flows into a container or through a hose to a nearby drain. Some dehumidifiers employ a shutoff switch that is operated by the weight of the collected water in the container when full, so that the unit automatically shuts down if left unattended.

Evaporator coils are generally constructed of either copper or aluminum. Steel is not used, since the coils are always exposed to moisture and would rust. Any leak which may occur in copper tubing is readily repaired by silver brazing. Aluminum coils are more difficult to repair, but can be done successfully by either brazing or by using aluminum epoxy compounds which are sold for this purpose.

Since the evaporator return line is usually at a temperature which is below the dew point, moisture will condense on it just as it does in the coil assembly. Most units will have the return line covered with insulation to avoid the problem of water falling from the tubing into the cabinet of the unit.

8.7 Fan Motor

The fan motor and blade are used to force air through the evaporator and over the condenser coils so that the dehumidification process may take place. These low-power motors usually operate for many years before they require service. Many fan motors used in dehumidifiers are lubricated for life, and do not have any provision for adding oil to the bearings.

When such motors fail because of dry or frozen bearings, it may be possible to coax some oil into the bearings along the shaft, thus restoring lubrication and adding service life to the motor. Non-detergent SAE 20 weight oil should be used. Sometimes complete disassembly of the motor will permit the bearings to be properly lubricated. If attempts at lubrication do not succeed, replacement is necessary.

Burnout of the fan motor stator winding is rare, and may be detected by a check of the resistance of the winding after disconnection of the leads from the dehumidifier electrical circuit. Burned out motors are not economically repairable; replacement is required.

8.8 Humidistat

Most dehumidifiers contain a humidistat (Figure 8.4), which is an automatic device that responds to relative humidity and controls the operation of a built-in switch. Often, the humidistat controls all functions of the unit, doubling as a power switch. The control is usually front panel mounted to allow the user to select the level of relative humidity that is desired. However, the accuracy of such controls is always in question, and any control which purports to control relative humidity levels below 50% may never be able to cycle the unit at such a level, since a dehumidifier's performance falls off sharply at low relative humidity levels.

Many humidistats operate by using a property, called hygroscopic, of a human hair or other organic material that causes its length to change with varying levels of relative humidity. The sensing material will

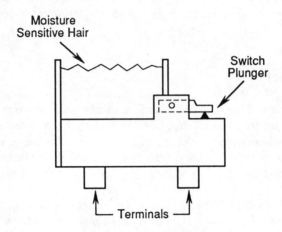

**Figure 8.4   Humidistat Operating by Change of Length
of Humidity Sensitive Hair**

shrink as the humidity level decreases, and stretch when it increases. A
spring-loaded mechanism is used to take up the slack as the hygroscopic
element stretches, and a switch is operated. This operates the dehumidifier compressor and fan motor.

Such humidistats may not be very accurate in their calibration.
However, a working unit will be able to turn the unit on and off as it is
rotated over its adjustment range. Many humidistats have a "constant
run" position at one end of the adjustment range. This permits the
humidistat control to be taken out of the circuit so that the dehumidifier
operates constantly.

8.9 Diagnostic Procedure

A dehumidifier, being a relatively simple device, is not difficult to
troubleshoot when inoperative. The first check is to verify that both the
fan motor and compressor are operating when power is applied. If either
is not, any electrical or mechanical problem associated with the inoperative
component must be corrected first.

Always disconnect the line cord from the power receptacle before attempting to disassemble or work on the dehumidifier. The parts and wiring inside represent a shock hazard.

Disassembly of the unit is usually very straightforward, and is usually done by removing all external screws that hold the cabinet in place. Some designs, however, are such that the humidistat and power control is mounted to the front of the cabinet assembly, which cannot be completely separated from the unit. This makes servicing a little more difficult, but not impossible.

Operation of the fan is easily checked by simply turning the unit on and observing if it quickly comes up to proper speed. If it seems sluggish, it may have dry bearings. The fan blade should free wheel with no apparent friction when the power is turned off and the blade spun by hand. Any motor which exhibits noticeable shaft friction will have to be either lubricated or replaced.

A motor which has a free running shaft but does not operate at all may have a burned out winding. Power to the motor may be verified by measurement of the voltage being applied to the winding, or its continuity may be checked after disconnection of the motor leads from the wiring of the dehumidifier. A normal reading will possibly be 100 ohms or more.

Lack of power to the fan motor and compressor may be caused by a defective humidistat or power control. Using a jumper wire to short out the switching contacts of the humidistat or power switch will verify if this component is the source of the problem.

It is sometimes very difficult to determine if a compressor is operating by listening to its normal operating sound, which is masked by the noise generated by the fan blade. One way to determine if the compressor is operating is to connect an ammeter in series with the line feeding the unit, or to use a clamp-on ammeter and observe the starting and running current of the compressor as the unit is switched on.

A normal unit will exhibit a large surge of current, for less than one second, that will settle down to the unit's nameplate rating. If the line current is zero, the wiring to the unit from the power plug, power switch, humidistat, compressor, and fan motor should be checked. The compressor overload, and compressor windings, should also be checked for continuity.

If the surge of current seems to last more than 1/2 second and then goes to near zero, the compressor did not start and its overload has opened the circuit. If power is left on, the cycle will repeat itself when the overload cools down.

A compressor that does not run may be caused by a stuck rotor, which is a non-repairable condition if caused by a mechanical problem within the hermetic unit. System pressures between the suction and discharge ports of the compressor which have not equalized can also result in a stuck compressor, but this symptom is caused by a problem in the refrigerant circuit, and not the compressor. A mechanically stuck compressor, or one with an open or shorted stator winding, is not repairable and must be replaced. Compressor starting problems may also be caused by a defective starting relay, capacitor (if used), or bad connection in the start circuit.

Before condemning a compressor as defective, the only way to be absolutely sure that it is defective is to remove all wiring from its terminals and substitute an external test circuit consisting of a line cord and new starting relay and overload. If a capacitor is part of the original circuit, this part should also be substituted with a known good one.

A starting relay may be simulated in those units which do not use capacitors by using a jumper lead to momentarily short out the start and run winding of the compressor upon power-up. This is illustrated in Figure 8.5. This simulates the action of the start relay, which normally makes momentary contact to the start winding when power is first applied to the compressor. By using an external line cord and clip lead to simulate the starting circuit, the service technician can determine if the compressor is indeed stuck.

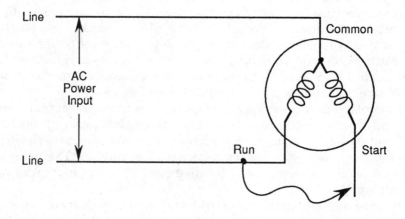

Figure 8.5   Manual Start Circuit using Clip Lead

A dehumidifier that has a normally running compressor and fan motor, but does not exhibit a cold evaporator coil, is probably very low on refrigerant. Rarely will the service technician encounter such problems as a compressor that does not pump, or a clogged capillary tube. A unit which develops frost on part of its evaporator will probably be low on refrigerant. Dehumidifiers generally use such a small quantity of refrigerant that virtually any loss may result in improper operation.

Any unit which develops frost should be left running for an hour in a room with an ambient temperature of 70 degrees F (21 degrees C) or more. If the frostup is sustained, the unit requires recharging. If the evaporator is wet throughout after the system is stabilized, the unit is probably not defective.

Dehumidifiers are constructed with completely sealed systems, and an access valve must be added to the compressor suction line, or process tube, to evacuate and charge a unit that has never seen service. A piercing type valve is the most convenient to use.

When the valve is connected to the system, the low-pressure manifold gauge should be connected to determine the existing residual pressure of the system. This is important, for it will provide essential data on the condition of the sealed system.

A dehumidifier that has some residual pressure does not have a catastrophic leak, and if no oil stains are evident on any part of the sealed system, a check with a detector will probably not reveal a leak. Such a system may be evacuated and recharged without further service.

If the pressure gauge indicates zero charge, the system may be then pressurized with refrigerant and checked for leaks. If necessary, the test may be facilitated by increasing the pressure up to 150 PSI, no higher, by adding nitrogen or carbon dioxide to the system.

A thorough leak test may still not reveal a problem. The loss of the original refrigerant charge may have taken many years, and if so no leak may be found. The system may be evacuated and recharged.

If the system seems to have a reasonable charge of refrigerant as indicated by a static pressure of 50 PSI or more, the compressor may be operated to ascertain that the compressor does pump the refrigerant through the system. This will be indicated by the pressure gauge reading which will decrease to near zero or below. If there is no change in pressure reading when the compressor operates, it is defective and must be replaced.

If the pressure reading goes into vacuum and there is no sustained heat developed in the condenser coils, the capillary is probably blocked. Since the suction inlet of the compressor is starved of refrigerant, it is not able to compress the gas and pump it to the condenser where its heat can be felt by hand. Another way to determine if the capillary tube is blocked in a charged unit is to operate the unit for a few minutes and turn it off. The sound of liquid refrigerant flowing into the evaporator should be heard. If the blockage is caused by moisture in the system which has frozen at the capillary tube outlet, the sound of refrigerant will be heard when the ice melts.

A clogged capillary may be replaced, but if this is done it is recommended that a small filter/drier be placed in the liquid line between the condenser and capillary tube to ensure that the new part will not become clogged. Be sure to observe the correct orientation of the filter/drier so that refrigerant flow is in the correct direction. Any system which contains a large amount of solid contaminants may have a compressor which is breaking down, clogging the filter. Such a condition must be investigated.

8.10 Repairs

Leaking evaporator or condenser assemblies, or their associated tubing, are usually easily repairable when made of copper. Aluminium tubing or components may be difficult to work with, but an experienced service technician should be able to successfully braze aluminum tubing using the correct materials. As with any brazing operation, cleanliness of the parts to be brazed is important; a good leak-tight joint cannot be produced if the parts are covered with dirt, corrosion, etc.

An aluminum epoxy, designed for the repair of leaks in aluminum evaporators and condensers, is available. These products are relatively easy to use. After any leak repair, the unit should be pressurized to at least 50 PSI with refrigerant, to check the repaired section for a leak. A better check would be to increase the pressure of the refrigerant with nitrogen, up to 150 PSI, for leak detection. The pressure should never be allowed to exceed this amount; the sealed system is not capable of withstanding higher pressures and may burst.

Some evaporators and condensers may not be repairable due to a condition of porosity which extends over a large area of the component. Replacement parts are usually available, at reasonable cost, from the manufacturer of the unit.

Unless there has been excessive oil loss from the system due to a catastrophic leak, it is usually not necessary to add oil to the system. Oil charge information is available from the manufacturer of the compressor or dehumidifier. If oil must be added to a system, be sure to use the correct grade of oil as specified by the manufacturer of the unit.

If the repair includes brazing of the capillary tube, it is important that the small ID of the tube is not covered or restricted in any way. A capillary may be successfully brazed to a tube of larger diameter by crimping the periphery of the large tube around the capillary in such a way as to create a U-shaped section, as illustrated in Figure 8.6. The brazing operation will seal the assembly.

Note that the capillary is inserted into the larger tube at least 3/4 inch (2 cm). This ensures that any flow of the molten brazing material does not reach the end of the capillary tube where it may close or restrict the opening. Flux should be applied to the parts in such a way as to preclude its entering the sealed system during the brazing operation.

After the brazed joint has cooled, any residual flux should be washed away with water to prevent corrosion, and to clear the assembled parts for leak detection.

Electrical replacement parts are available from the manufacturer of the unit, or a distributor of refrigeration and air conditioning supplies. It is not necessary to obtain exact replacements for the humidistat, fan motor, or compressor. If a substitution of the compressor is necessary, obtain one which has the same voltage and current rating of the original. The compressor must be rated for high back pressure operation using R-12 refrigerant (or any other recommended substitute).

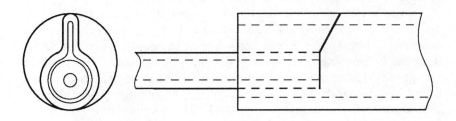

Figure 8.6   Large Tube Crimped Around Small
Tube Prior to Brazing Operation

When a compressor is replaced, the electrical accessories such as the starting relay, overload, and capacitor (if used) must be matched to the new part. Do not use the original parts if the new compressor is not the exact same unit. To do otherwise may result in damage to the new compressor.

New compressors are usually charged with the correct amount of oil. If in doubt, check with the source from which the compressor was obtained. Lack of proper oil charge will result in a permanently damaged compressor.

8.11 Evacuation and Charging

Total evacuation of a dehumidifier is mandatory before recharging. There is always the possibility that some moisture has seeped into the system through the same opening(s) in which the original refrigerant has escaped. A dehumidifier which is operated with a low refrigerant charge will almost certainly have a compressor suction pressure of less than zero (vacuum), and under this condition the humid air surrounding the unit during such operation can enter the system. If evacuation is not performed prior to charging, the unit may not operate properly and could suffer internal damage due to moisture in the system.

Figure 8.7 illustrates an evacuation/charging setup. Note that the dehumidifier should be charged at an ambient temperature of 65 degrees F (18 degrees C) or more. Operation at lower temperatures may cause frost buildup on the coils, which is a normal condition for some units operating at too low an ambient temperature level.

The two valves of the manifold gauge set are opened, and the refrigerant (usually R-12) container valve is closed. If the access valve on the dehumidifier suction line does not automatically open when the charging hose is connected, this valve must also be opened. Note that the refrigerant container is kept in an upright position so that gas, and not liquid, is discharged into the system during the charging procedure.

Power to the dehumidifier is turned off. The vacuum pump is operated and the low-pressure gauge observed to monitor the evacuation process, which may take a few minutes to more than an hour, depending upon the displacement of the pump and amount of moisture in the dehumidifier system.

The pump should be capable of reaching a vacuum level of at least 29.5 inches of mercury (about 100 KPa) at sea level, and the pump operated until its maximum capability is reached. At altitudes above sea level, the

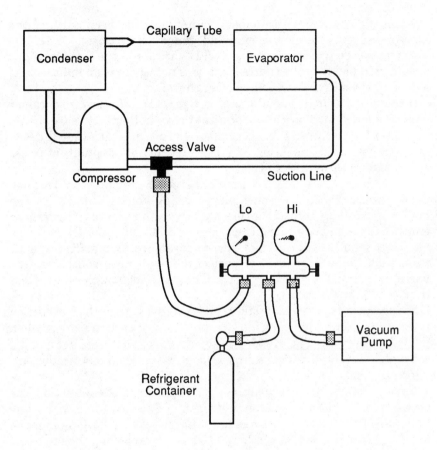

Figure 8.7   Evacuation/Recharging Setup

maximum attainable vacuum is reduced about 1 inch of mercury for each 1000 feet (305 meters). For example, at an elevation of 1000 feet the system should be pumped down to 28.5 inches (about 96 KPa) of vacuum.

If the level of vacuum seems to reach a plateau at some point below maximum pump capability, the pump is probably removing water from the system as it is being vaporized at the lowered pressure level. The pump should be allowed to continue until all the moisture is removed, which will be indicated by a gauge reading equal to maximum pump capability.

When the maximum attainable vacuum level has been reached, the high-pressure valve on the manifold gauge set is closed. The low-pressure gauge will indicate the level of vacuum in the system, and it should hold for several minutes. If it does not, there is an indication of a leak which must be corrected before charging.

If the system holds vacuum and is tight, the refrigerant container valve is opened to allow gas to enter the dehumidifier. Pressure should be allowed to rise until it is about 30 or 40 PSI, at which time the low-pressure gauge valve is closed. This is to prevent any possibility of excess refrigerant entering the system.

The dehumidifier is now powered to operate the compressor. This will be indicated by the low-pressure gauge reading which will decrease. The low pressure gauge valve is opened slightly to allow more refrigerant to enter the system as the compressor draws it in. Periodically the valve is closed to allow the system suction pressure to stabilize and be measured. Frost may form on part of the evaporator coils, since the partially charged system may be operating at a suction pressure of less than 30 PSI at this time.

Refrigerant is allowed to flow into the system until the suction pressure is about 40 PSI, as indicated by the low pressure gauge when the valve is closed. The evaporator should be cold in its entirety, and if the ambient humidity level is high enough water should be accumulating on the coils.

During the charging procedure, the temperature of the suction line between the evaporator and compressor should be monitored. If this line becomes very cold, the system is fully charged. Overcharging must be avoided, since it can result in liquid refrigerant reaching the inlet port of the compressor and cause damage. It also results in a higher than normal evaporator pressure and reduced dehumidification capability.

The charging process may be considered to be complete when the entire evaporator is cold and the compressor suction pressure is at least 40 PSI. Note that at high ambient temperature levels the normal suction pressure of the system will be higher. If possible, obtain a spec sheet on the model being serviced to determine the manufacturer's specifications on normal suction pressure at various ambient temperature levels.

Before disconnecting the charging hoses, the refrigerant container valve and dehumidifier access valve (if not spring loaded) must be closed. A cap should be placed on the access valve to prevent any dirt from entering, and to ensure a leak-tight valve. The leak detector should be used to verify that the access valve is indeed tight.

Heat Pumps

9.1 General Information

Heat pumps are a specialized form of compression cycle air conditioning units which are designed to transfer heat in either of two directions. During the cooling season, the heat pump acts as a common air conditioning system, removing heat from the interior of a structure and transferring it to the outside. In winter the role of the condenser and evaporator sections of the unit is reversed so that the heat produced by compression of the gaseous refrigerant is released to the interior of the structure. Reversal of the heating and cooling operation of the heat pump is accomplished through the use of a reversing valve, which directs the hot, compressed refrigerant produced at the compressor discharge port to the inside or outside coil as required.

The heat energy which is delivered to the interior area during cool weather must come from the outside air. Although this air may seem to be "cold" during wintertime, it contains sufficient heat energy which can be captured by the heat pump and delivered to the interior of a building.

Heat pumps are rated by several standards. One is the coefficient of performance (COP), which is simply the ratio of the electrical power input in BTU/hr. divided by the BTU output in BTU/hr. The BTU input of the system is simply the wattage consumed by the unit multiplied by the factor 3.413. The COP of a heat pump will vary with outside temperature since the amount of available heat contained in the outside air falls with lower temperature. Calculations of COP have been standardized by the Air Conditioning and Refrigeration Institute (ARI) at two outside air temperatures: 47 degrees F (8.3 degrees C) and 17 degrees F (–8.3 degrees C).

The Heating Seasonal Performance Factor, HSPF, is a Department of Energy performance rating which is calculated by dividing the seasonal BTU output of a heat pump by its wattage input, using a national average outdoor temperature during the heating season. The HSPF includes power consumed during defrost cycles, and during operation of supplemental heat.

A very popular heat pump rating factor is SEER, Seasonal Energy Efficiency Ratio. This ratio is calculated with the heat pump operating in cooling mode, with an outside air temperature of 82 degrees F (27.8 degrees C). The calculation is the same as EER, which is BTU/hr. divided by watts input. EER is calculated at an outside air temperature of 95 degrees F (35 degrees C).

Heat pumps may be designed as packaged or split systems, as are air conditioning units. It is also possible for a heat pump system to be designed as "triple split" (Figure 9.1), which allows the compressor and reversing valve to be located inside the building at a location remote from the inside coil.

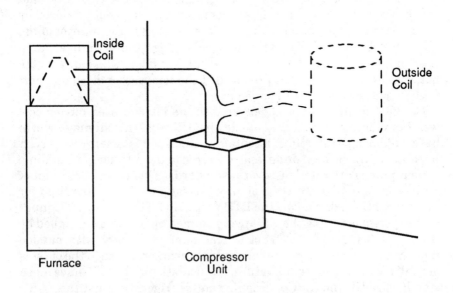

Figure 9.1   Triple-split Heat Pump System.
(Courtesy of Carrier Corporation)

9.2 System Capacities

The capacity of a heat pump system is specified at two outside air temperatures, as required by the American Refrigeration Institute (ARI). These temperatures are 47 degrees F (8.3 degrees C) and 17 degrees F (–8.3 degrees C). The reason for this is the varying available heat output of a heat pump, which varies inversely with the outside temperature. Additionally, heat pump specifications take into account the losses which are incurred in the defrost cycle. Manufacturers supply BTU output data for each model of heat pump that is produced, in accordance with the prevailing outside air temperature. Typical heat pump output capability is illustrated in Table 9.1.

9.3 Balance Point

The concept of balance point is an important factor which should be understood. It is defined as the outdoor temperature at which the output of a heat pump is equal to the heat loss of the building. Since these two quantities vary in an inverse manner with each other as outside air temperature changes, the point at which they are equal is called the balance point of the system. Outside air temperatures which are below the balance point will result in the heat pump system operating 100% of the time. The use of supplemental heat to maintain the desired interior temperature is required at outside temperatures below the balance point. The supplemental heating system will cycle on and off, as controlled by the second stage of the thermostat.

9.4 Compressor

Although heat pumps seem to be very similar to compression cycle air conditioners, they are subject to very special operating conditions which necessitate important design differences. One of these is in compressor design.

During the cooling season, an air conditioning compressor sees a compression or pumping ratio of only 3 to 1. However, when outside air temperature falls below 40 degrees F (4.4 degrees C), the compression ratio increases. This is one reason why compressors rated for air conditioning service only should never be used in heat pump systems. To do so will invite failure of the compressor.

Heat Pump Integrated Heating Capacities

Nominal Unit BTUH	Temperature of Outdoor Air Entering Unit								
	-10F	0F	10F	20F	30F	40F	50F	60F	
	-23C	-18C	-12C	-7C	-1C	4C	10C	16C	
	Capacity (BTUH)								
15,000	5,000	6,000	7,000	9,000	10,000	12,500	15,000	17,500
18,000	5,000	6,500	9,000	11,500	14,000	16,500	19,000	21,000
24,000	6,500	9,000	12,000	15,000	19,000	22,500	27,000	32,000
30,000	9,000	12,000	15,500	19,000	22,500	27,000	32,000	37,500
36,000	12,000	16,000	20,000	24,000	29,000	34,000	41,000	48,000
42,000	14,000	18,500	23,500	28,500	34,000	41,000	48,000	55,000
48,000	14,000	18,500	24,000	29,500	35,500	42,000	49,000	57,500
60,000	19,000	25,500	32,000	38,500	45,000	52,000	59,000	67,500

Table 9.1   Typical Heat Pump Capacities as a
Function of Outside Air Temperature
(Courtesy of Carrier Corporation)

Heat pump compressors are designed to operate with a compression ratio of 5 to 1 at outside air temperatures of –10 degrees F (–23 degrees C). They are designed with larger bearing surfaces and heavier connecting rods and wrist pins to withstand the higher stresses involved. The motors are designed with heavier insulation to withstand the higher operating temperatures. Even the pistons are designed to accommodate the changing cylinder wall clearance that occurs with expansion and contraction at temperature extremes found during defrost cycles.

The use of a crankcase heater for the compressor is mandatory in a heat pump application, since the cold ambient outside air temperatures will, during idle periods, cause the refrigerant to migrate and collect in the sump of the compressor. This liquid refrigerant will dilute the lubricating oil, and possibly be drawn into the suction intake of the compressor upon start-up. These conditions will cause damage and premature failure of the compressor.

Crankcase heaters are electrically operated, and consist of a strap-around resistance element which is placed around the shell of the compressor, or it may be a self-regulating immersion type in a well that is part of its internal construction. An alternate method of maintaining compressor temperature during idle periods is to allow a trickle current to pass through the start capacitor of the compressor, which heats up the start winding sufficiently to prevent the collection of liquid refrigerant in the sump. This method is not as efficient as the prior two methods.

9.5 Reversing Valve

The reversing valve (Figure 9.2) is the one component in a heat pump system which distinguishes it from an ordinary air conditioner. The purpose of this valve is to direct the hot, high-pressure discharge gas from the compressor to either the outside coil for cooling mode, or to the inside coil for heating. Most reversing valves have four ports, although it is possible for a manufacturer to design a heat pump system with two three-port valves. The valve ports are connected to the compressor suction and discharge lines, and the inside and outside coils, to direct the refrigerant to and from the desired coil.

In a typical reversing valve, three ports are contained on one side of a barrel, and the fourth port on the other side. The indoor and outdoor coils are connected to the outside ports in the group of three, with the center port connected to the suction inlet of the compressor. The remaining port is connected to the discharge line of the compressor.

The reversing valve contains a main valve assembly and a pilot valve assembly. An electrically operated solenoid coil acts upon the spring loaded pilot valve, which controls a bleed system that causes an appropriate response of the main valve.

The pilot valve bleed system is connected to each end of the main valve assembly, and the suction line of the system, so that the low pressure produced at the compressor suction inlet can be impressed upon either side of the main valve assembly.

When the solenoid is not energized, the spring-loaded piston of the pilot valve remains in its dormant position. This allows the discharge gas of the compressor to bleed through a small passageway, filling a large cavity on the left of the main valve. This causes the main piston to move to the right and close off the tube port to the low side of the system. This

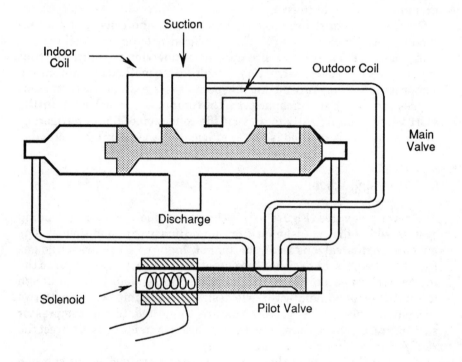

Figure 9.2   Typical Reversing Valve
(Courtesy of York International Corporation)

position of the main valve, with the solenoid disengaged, directs the hot discharge gas from the compressor to the inside coil for heating mode.

When the solenoid coil is energized, the pilot valve plunger is pulled against the spring, allowing the high-pressure gas in the main valve to escape through the pilot valve capillary tubes to the suction line of the compressor. This allows the main valve to move to the left, pushed by discharge gas pressure. At the end of its travel the bleed port on the left side of the reversing valve is closed off, and the reversing valve is in the cooling mode.

Note that the reversing valve is designed so that if for any reason the solenoid is not engaged, the fail-safe position of the valve is in the heating mode. This avoids any possibility that the system would remain in cooling mode due to failure of the solenoid circuit, which would be an unacceptable condition during the heating season.

Problems with reversing valves may show up in both modes of operation. If the valve does not fully shift from one side to the other, the bleed-down capillary on each side of the valve will be hot to the touch due to the flow of discharge gas. The system suction pressure will be higher than normal, and discharge pressure lower than normal, due to the intermixing of discharge and suction gases.

9.6 Heating and Cooling Refrigerant Circuit

As with an air conditioning system, the flow of refrigerant must be metered into the inside coil to create a pressure differential between the inside and outside for cooling. This is done through the use of a capillary tube, expansion orifice, or an expansion valve. When the heat pump system is operated in heating mode and the role of evaporator and condenser interchanged, a second metering device is required for the outside coil. Figures 9.3 and 9.4 illustrate the differences in refrigerant flow for each mode of operation.

9.7 Check Valve

Although the heat pump refrigerant circuit employs two metering devices, only one must be in operation at a time, depending upon the operating mode of the system. Since the unused device must be taken out of the circuit to allow full refrigerant flow, a check valve is connected in parallel with each device. This permits refrigerant to flow in one

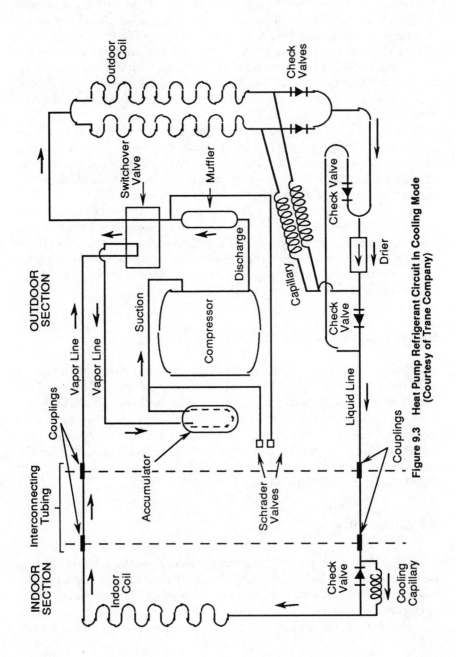

Figure 9.3   Heat Pump Refrigerant Circuit in Cooling Mode (Courtesy of Trane Company)

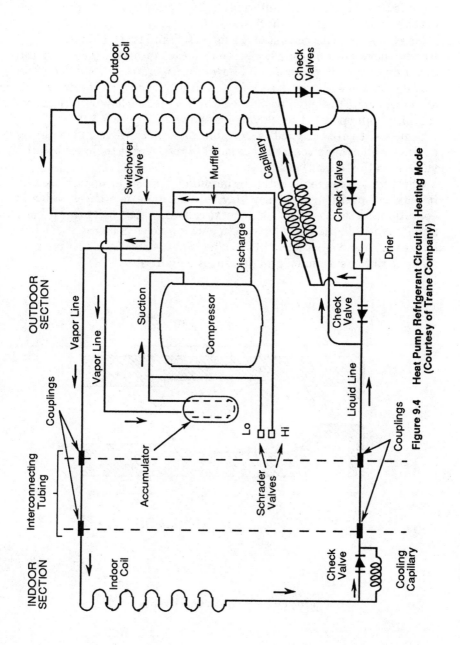

Figure 9.4 Heat Pump Refrigerant Circuit in Heating Mode (Courtesy of Trane Company)

direction without restriction, while in the opposite direction the check valve is closed by pressure differential. This allows the metering device to take control of refrigerant flow.

Many check valves consist of a magnetic stainless steel ball which is free to move within a brass retainer sleeve, in accordance with the direction of system pressure and flow of refrigerant. At one end of its travel the steel ball acts as a valve and shuts off the flow of refrigerant; at the other end the ball is positioned to allow the free flow of refrigerant. A typical ball-operated check valve is shown in Figure 9.5.

Some valves are designed as a combination check valve and metering device by allowing a controlled amount of refrigerant to flow when the valve is in the "closed" position.

Check valves can malfunction by becoming stuck in an open, closed, or in-between position. An easy way to determine if a check valve is operational is to use a strong magnet to cause the steel ball to move back and forth inside the valve body. If a clicking noise is heard as the ball strikes each end of its travel, the valve is not stuck. This test must be made with the system off and pressures equalized.

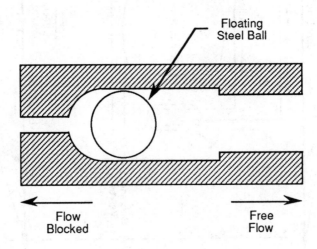

Figure 9.5   Typical Check Valve

A check valve that is stuck in one position will present system operating problems in only one mode of operation. Understanding correct refrigerant flow will allow determination of proper valve operation by noting the temperature difference between the input and output sides of the valve. For example, if the check valve is supposed to be open but is stuck closed, this will result in a temperature differential between each side of the valve, which can be felt by hand.

9.8 Accumulator

Heat pump systems employ an accumulator (Figure 9.6), placed in the compressor suction line, to store any liquid refrigerant that does not

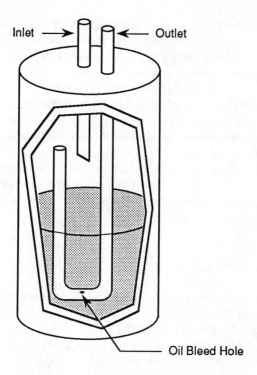

Figure 9.6   Suction Line Accumulator

vaporize in the cooling coil. When operating in cooling mode, virtually all of the liquid refrigerant metered into the inside coil will be vaporized, but when the system is operated in heating mode, the reduced heat load on the outside coil may not be sufficient to convert all the liquid refrigerant to gas.

In order to protect the compressor against any possible liquid refrigerant entering its suction port, the accumulator stores the liquid and releases it as a gas. Oil which travels through the system with the refrigerant is also collected in the accumulator, and a small bleed hole is placed in the accumulator return tube to allow the oil to return to the compressor with the gaseous refrigerant.

9.9 Filter/Drier

Some heat pump systems are equipped with liquid line filter/driers. When repair work is performed, it is recommended that a "solid core" filter/drier be installed. It is important to note that filter/driers used on heat pump systems must be able to handle refrigerant flow in both directions, if installed at a location where the direction of flow reverses in accordance with the heating and cooling modes of operation.

If a compressor burn-out has occurred, a solid core filter/drier should be installed in the suction line just before the replacement compressor, to protect it from any possible system contaminants. Pressure drop across the filter/drier can be monitored after the system is placed back in operation to determine if it becomes clogged from residual contaminants in the system. A pressure differential of 5 PSI indicates that the filter/drier needs to be replaced.

9.10 Defrost Cycle

In order for the heat pump system to extract heat energy from the outside air when in heating mode, the outside coil must be operated at a level which is at least 10 degrees F (5.6 degrees C) lower than the outside air temperature. This means that at most outside air temperatures during the heating season the coils will operate below the freezing level. As a result, ice may form on the coils. Figure 9.7 illustrates the propensity of a heat pump system to form frost on its outside coils.

The amount of ice buildup is a function of the outside air temperature. At temperatures above 45 degrees F (7.2 degrees C), not much ice will form because the coils will not be as cold. Most ice will form at outside air temperatures around the freezing mark. Below that, the air does not contain as much moisture and less frost will build up.

The formation of ice on the outside coils will inhibit air flow and insulate the tubing, so that insufficient heat transfer will take place. The heat pump system must periodically provide heat to the outside coils to melt away the accumulation of frost and restore normal system operation. This is usually accomplished by automatically reversing the heat pump system to cooling mode so that hot compressor discharge gas is passed through the outside coils to melt the accumulated ice. The outside fan motor is disabled during the defrost cycle. At this time, the system must provide supplemental heat, usually by means of electrical resistance heaters, to temper the cooled air generated in cooling mode and to help maintain the desired interior comfort level.

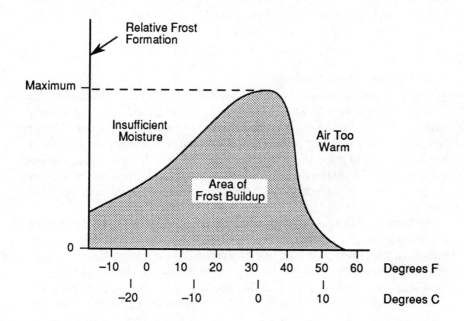

Figure 9.7   Relative Frost Formation Versus Outside Air Temperature (Courtesy of York International Corporation)

There are four methods by which heat pump systems initiate the defrost cycle. Each of these is designed to start the sequence when sufficient ice has formed, and to stop it as soon as the ice has melted away. Excessive defrosting time causes unnecessary loss of interior heat and increases electrical operating expenses. Defrost cycle time may be 3 minutes or more, depending upon the design of the system.

One method used to initiate the defrost cycle is to monitor the air pressure drop across the outside coil and detect when the formation of ice causes restriction to the flow of air. Another method is to use a timed cycle, much like a frostless refrigerator, and initiate the defrost function at periodic intervals providing that an outside temperature sensor indicates that the coil is cold enough to form ice. A modification of this method uses solid state circuitry to monitor the outside air temperature and vary the timing of the defrost cycle accordingly.

The most sophisticated defrost cycle control uses a microprocessor (solid state computer circuit), which determines when to initiate defrosting, based on information provided by sensors that monitor outdoor temperature and liquid line temperature. This system is called the Adaptive Delta-T Demand Defrost cycle control.

9.11 Supplemental Heaters

Heat pump systems used in many areas of the country cannot supply sufficient heat during normal cold temperature extremes. Outside air temperatures which fall below the balance point of the system will cause the heat pump to operate continuously, and also will require supplemental heat to maintain the desired indoor comfort level. This heat is usually supplied by built-in electrical heaters, but also may be provided by fossil fuel or solar sources.

For best operating efficiency, electrical supplemental heat is provided by several stages rather than one high-capacity heater. When the outside air temperature falls just below the balance point, the first stage of electrical heat is initiated. Should the operation of the heat pump and first stage of supplemental heat be insufficient to provide the necessary comfort level, a second stage of supplemental heat cuts in. This process may continue with a third stage, if necessary. By using several stages, only the minimum amount of supplemental heat needs to be used to provide the necessary comfort level.

9.12 Emergency Operation

Some manufacturers of heat pump systems include an "emergency heat" switch position in the thermostat control circuit. An indicator light may be employed to alert the user that the heat pump system has a malfunction, and the system should be switched over to supplemental heat mode until a service technician can restore normal operation.

When operating under emergency heat conditions, the heat pump system is disabled and all heating is provided by the supplemental heaters of the unit. Thermostatic control is still maintained, but the loss of heat energy provided by the heat pump may result in indoor temperatures which fall below the thermostatic setting.

9.13 Solid State Control Modules

One of the most advanced solid state control modules used in heat pump systems is the Yorkguard module, pioneered by York International Corporation, York, Pennsylvania. This is a microprocessor controlled solid state control circuit which automatically handles all necessary and various operating modes of the heat pump system. It also includes self-diagnostics, which, upon system malfunction, are able to alert the service technician of the area of fault in the heat pump system. This makes servicing easier.

The Yorkguard module provides an automatic 5 minute off cycle when the compressor shuts down, either by thermostat control or momentary power loss. It monitors the line voltage feeding the system and prevents operation if the supply voltage falls below a predetermined minimum value (commonly called brown-out protection).

The module allows field adjustment of the balance point of the system, over a range of 0 to 45 degrees F (–18 to 7 degrees C), so that the supplemental heating system can be locked out at outside temperatures above the balance point setting. This feature ensures minimum electrical operating cost.

The module has built-in logic for fossil fuel "add-on" systems. It permits field adjustment of the low ambient cut-off temperature at which the heat pump is shut down, when outside air temperatures fall too low for economical cost of electrical operation versus fossil fuel cost. The low ambient cut-off temperature may be set to within the range of –10 to 45 degrees F (–23 to 7 degrees C).

**Table 9.2   Yorkguard Diagnostics Failure Code
(Courtesy of York International Corporation)**

Fault Code	Possible Cause
2	Discharge pressure reaches 400 PSI
3	Discharge temperature reaches 275 F (135 C)
4	Discharge temperature does not reach 90 F (32 C) within 1 hour of operation
5	Two default defrosts occurred within 1 hour
6	No fault specified
7	Outdoor sensor failure
8	Liquid line sensor failure
9	Bonnet sensor shorted

The Yorkguard module also controls the defrost cycle of the system, using the Adaptive Delta-T Demand Defrost technique.

Built-in fault diagnostics (Table 9.2) is implemented through the use of a blinking light emitting diode (LED), which provides a code number from 2 to 9 that alerts the service technician of the area of fault. This can greatly facilitate the servicing process.

9.14 Performance Chart

In order to determine if a heat pump system is not operating up to specification, it may be necessary to measure system pressures and compare the readings to the data that is supplied by the manufacturer of the unit. Such data is available in service bulletins. Pressure measurement data is supplied in both heating and cooling mode. If a unit does not exhibit the expected system pressures, service is required. This may be as simple as adding refrigerant, but could also be caused by malfunction of a component. A typical pressure/temperature chart is illustrated in Table 9.3.

Table 9.3   Typical Pressure/Temperature Performance Chart
(Courtesy of Lennox Industries, Inc.)

Mode & Outdoor Temperature			Discharge Pressure	Suction Pressure
Cooling	75 F	24 C	180 PSI	75 PSI
Cooling	85 F	29 C	209 PSI	77 PSI
Cooling	95 F	35 C	238 PSI	79 PSI
Cooling	105 F	41 C	270 PSI	81 PSI
Heating	20 F	-7 C	175 PSI	33 PSI
Heating	30 F	-1 C	188 PSI	42 PSI
Heating	40 F	4 C	201 PSI	51 PSI
Heating	50 F	12 C	214 PSI	61 PSI

9.15 Evacuation and Charging

Whenever any component has been replaced, or the sealed refrigerant system repaired or opened up to the atmosphere, the heat pump must be evacuated to at least 500 microns. Heat pumps are especially sensitive to residual moisture because they operate at very low evaporator temperatures in heating mode. Any small amount of moisture in the system can turn to ice and show up as a restriction at the metering device. Because of this, a deeper vacuum than that required by an air conditioner is necessary.

One method to remove moisture from a system is to evacuate it to the maximum level obtainable from the vacuum pump, and then to break the vacuum to zero pressure by adding dry nitrogen to the system. A second evacuation to remove the nitrogen, and any moisture that it has trapped, is then performed. The system may then be charged to the correct level with refrigerant.

By far the best method to charge a system with the correct amount of refrigerant is to charge it with the exact amount, by weight, as specified by the manufacturer. If this information is not available, the service technician must rely on other methods to determine full refrigerant charge.

Charging a heat pump system is best done in cooling mode, if possible. If indoor air temperature is below 74 degrees F (23 degrees C) the supplemental heating system can be operated until interior temperature is raised to that level. This will create the necessary load for the indoor coil when operating the heat pump in cooling mode.

The thermostat can then be set to 68 degrees F (20 degrees C) and the unit operated after charging until system pressures stabilize. If outside air temperature is below 60 degrees F (16 degrees C) the air flow through the outside coil must be restricted to bring discharge pressure to within the 200 to 250 PSI range.

The liquid line temperature is measured with a thermometer. The outside air temperature is also measured, and the difference (liquid line temperature minus air temperature) should be approximately 7 degrees F (4 degrees C). Refrigerant may be added to the system to approach this temperature difference.

An alternate method of determining the approximate necessary refrigerant charge is called the subcooling method. This is done by comparing the liquid line temperature to the condensing temperature as indicated on the temperature scale of the high-pressure gauge. The difference between the two (condensing temperature minus liquid line temperature) is the subcooling temperature, and should be between 7 and 12 degrees F (4 and 7 degrees C). Refrigerant can be added to produce this amount of subcooling.

The only way to "correctly" charge a heat pump is by evacuation and adding the required amount of refrigerant, by weight, as specified by the manufacturer of the system. This method ensures that any system malfunction will be evident to the service technician, and adding or subtracting refrigerant charge is not the solution to the problem. If a known correct charge is added to a system and the operating pressures are not right, analysis of the wrong pressure readings should lead directly to the problem area.

Using the pressure charts supplied by the manufacturer is an alternate method of determining correct refrigerant charge in a system which is operating properly, but there is always the possibility that the service technician will attempt to compensate for problems by adjusting the amount of charge.

9.16 Troubleshooting

An experienced air conditioning service technician should be able to diagnose and repair heat pumps. Components which are unique to heat pump systems have been covered in this chapter, and any problem which develops will most likely affect both heating and cooling modes. An air conditioning technician will probably prefer to check a heat pump under the cooling mode of operation first, since he or she will feel more familiar with the system.

There will come a time when the service technician is called upon to repair a unit during the cooling season, when outside air temperatures may be too low to operate the unit in cooling mode. However, by using the information provided by sight, sound, and feel, the probable cause of a fault may be readily determined even without the need to connect the manifold gauge set.

As with an air conditioning system, the unit should be checked first for clean filters, free air flow through indoor and outdoor coils, tight belts, properly lubricated blower motors, and adequate line voltage at the power input terminals. Condensate drainage on both indoor and outdoor coils should be inspected for proper operation, including checking condensate pumps if used. Any of the above conditions, if not correct, may cause less than satisfactory performance of a heat pump system.

When servicing a heat pump system, it should be checked in both heating and cooling modes. Heat pumps automatically switch to cooling mode during the defrost cycle, and improper operation in either mode can prevent the unit from operating normally during the heating season. Proper operation in one mode and not the other can help lead directly to the fault.

Many problems associated with heat pumps will be electrical in nature. These systems utilize far more complicated electrical control circuits than do air conditioners, as illustrated in Figure 9.8, and it would be prudent to first obtain a service manual from the manufacturer or distributor of the system before attempting to troubleshoot an electrical problem. It may be extremely difficult to locate an electrical fault in a relatively complex heat pump control circuit without a schematic diagram of the circuit.

Once the fault has been located and repaired the unit should be operated in both heating and cooling modes. If work has been done on the sealed refrigeration system, the operating pressures can be compared to the pressure/temperature data provided by the manufacturer of the unit. When the unit is operated in the heating mode, sufficient operating time should be allowed to observe the defrost cycle if outside air temperatures permit it.

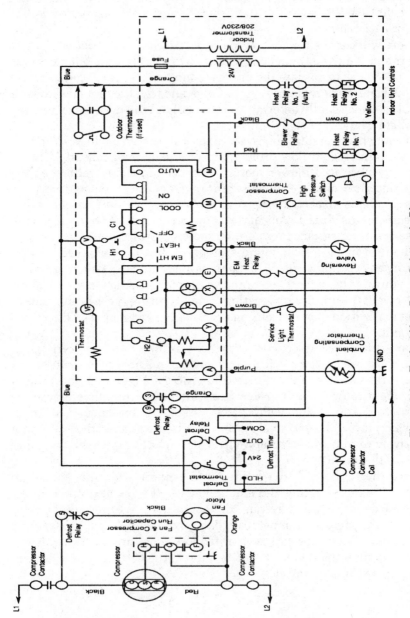

Figure 9.8   Typical Heat Pump Schematic Diagram (Courtesy of Lennox Industries, Inc.)

10

Ammonia Absorption Systems

10.1 General Information

The ammonia absorption system of air conditioning is a viable alternative for residential and commercial use, and offers several advantages over the compression cycle system. Depending upon the relative operating cost difference between gas and electricity, the cost of running an absorption air conditioning system may be significantly lower. Since there is no compressor in this type of system, its ambient noise level will usually be less. These systems are becoming more and more popular, and an air conditioning service technician will eventually be called upon to service them.

An absorption system has certain similarities to the compression system, but there are many differences which must be understood. These systems have more complicated components and controls, and analyzing faults is somewhat more difficult.

It is highly recommended that you obtain the service manual for the particular system under repair. These publications go into far more detail than can be provided in this chapter. Many systems are supplied with service instructions which are given to the user at the time of installation, and the service and maintenance information contained in such manuals is very useful.

Additionally, each manufacturer of absorption systems has components, controls, and operating characteristics which may be different from each other. The information provided in this chapter pertains to the ammonia absorption air conditioning system manufactured by The

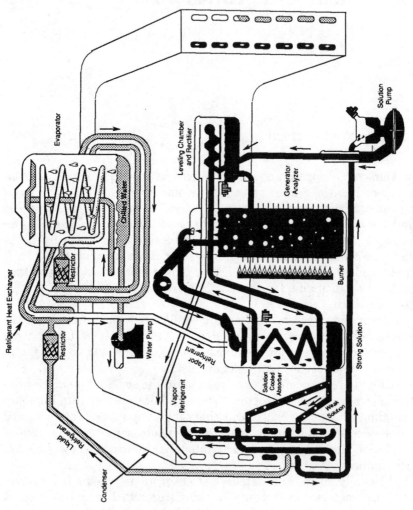

Figure 10.1   Absorption System

Dometic Corporation, Evansville, Indiana, under the Servel brand name, and will provide some insight and information on the troubleshooting and repair of the Servel gas fired air conditioner. Other manufacturers' units use similar components and the techniques described here can be used on those units as well.

10.2 Absorption Cycle

A typical ammonia absorption air conditioning system consists of an outside unit, called a chiller, and an inside unit called an air handler. The chiller contains all the necessary components which are required to provide the absorption cycle of refrigeration. Heat transfer between the inside and outside parts of the system is accomplished by means of a circulating water system which picks up heat from the controlled area and delivers it to the chiller, where it is removed. Figure 10.1 is a pictorial diagram of a typical ammonia absorption system.

The chiller contains the generator, condenser, evaporator, absorber, solution pump, and other components which are part of the absorption cycle. A source of gas, either natural or propane, is used to provide the heat energy upon which the absorption system operates.

In this system, ammonia (refrigerant 717) is the refrigerant and water is the absorbent. The ratio between these two compounds is 2 parts of water to 1 part of ammonia. Cool water or a weak solution of ammonia and water, used as an absorbent, is capable of absorbing large quantities of ammonia gas from the evaporator section of the sealed refrigeration system, just as a compressor can draw refrigerant gas at its suction inlet.

The generator contains a strong solution of ammonia and water which is heated by the gas burner when the thermostat calls for cooling. The boiling solution causes ammonia, in the form of a gas, to be driven out under pressure where it flows to the condenser. The absorption system is a two-pressure system with the high-pressure side operating between 200 and 300 PSI, and the low-pressure side operating between 40 and 60 PSI. System pressures vary with outside ambient temperature, just as they do in a compression cycle air conditioning unit.

The high-pressure ammonia gas gives up its latent heat in the condenser, which is cooled by air flow provided by the condenser fan. The ammonia, now liquid, passes through a restricting device to lower its pressure and allow it to change back to gas as it absorbs heat in the evaporator.

The source of heat which vaporizes the liquid ammonia is provided by water, which is circulated between the chiller and air handler. When the water passes through the air handler unit it picks up heat from the conditioned area, resulting in a temperature increase of about 10 degrees F (5.6 degrees C). The water coolant is then circulated to the chiller where it gives up its heat to the evaporator. After passing through the heat exchanger in the chiller, the cooler water is circulated back to the air handler to absorb more heat and the cycle repeats.

Heat from the circulating water, absorbed by the liquid ammonia in the form of latent heat, changes the ammonia to gas. The gaseous ammonia then travels to the solution cooled absorber, which contains a hot, weak solution of water and ammonia. The solution is cooled by air flowing across the absorber. The ammonia gas is quickly absorbed, and the resulting strong ammonia solution is pumped back to the generator to continue the cycle. The solution pump must handle the ammonia solution, and operates by means of a pulsating diaphragm. A hydraulic pump is used to deliver the pulses of oil pressure, up to 400 PSI, which operate the diaphragm.

10.3 Tools and Equipment

Most tools that are required for servicing absorption air conditioning systems are normally part of any air conditioning service tool set. However, there is one distinction which must be observed with any manifold gauge set that is to be used on ammonia systems. Ammonia is corrosive to copper and brass, and only steel tools and instruments should be exposed to ammonia gas or solution. For this reason, the charging manifold and hoses should not contain any brass parts or fittings.

Additional tools required in servicing ammonia absorption systems include a set of thermometers for measuring temperature drop across the air handler, and also to measure the temperature drop of the circulating water as it passes through the chiller unit. The information provided by these temperature readings will help in diagnosing and adjusting a system, and in determining if proper operating conditions exist.

A manometer will be required to measure gas pressure at the burner. Such pressure must be held within specifications for proper system operation. For natural gas systems the manometer should be capable of indicating a pressure of at least 5 inches of water column. Propane-fired units will require measurements of at least 10 inches of water column.

A DC microammeter (0 to 50 uA) may be used to measure current in the ignition circuit in accordance with the service instructions supplied by the manufacturer of the air conditioning system. This measurement will help diagnose a defective ignition circuit.

An ordinary level should also be part of the service technician's equipment. This will be used to check the orientation of the chiller, which must be level for proper operation.

A purging bucket will be required when purging the sealed refrigeration system. It should have at least a 2-gallon (8-liter) capacity and be made out of plastic, preferably any color other than yellow.

10.4 Safety Precautions

When working on the sealed refrigeration system, a garden hose attached to a nearby faucet should be available for safety purposes and to wash away any possible spillage of ammonia solution. Such solution, called Aqua-Ammonia-Chromate, contains Ammonia (refrigerant 717), Sodium Chromate Tetrahydrate, Sodium Hydroxide, and distilled water. It must be treated with respect since it is corrosive and damaging to the human body.

Chemical goggles should be worn when working with the ammonia solution, which can cause permanent injury to the eyes. Avoid spillage on the skin. Wear rubber gloves and protective clothing. Work only in a well-ventilated area. Use fans if necessary.

Any solution that gets into the eyes should be flushed out with large quantities of water for at least 30 minutes. Seek medical attention.

If exposed to the solution, the skin should be washed with soap and water for 15 minutes. Contaminated clothing should be removed immediately and washed before reuse. Consult a physician if there is any indication of ammonia burns or skin irritation.

If exposed to ammonia, get to fresh air immediately. If ammonia gas has been inhaled and breathing stopped, give artificial respiration. Seek medical attention. Qualified personnel may administer oxygen. If the person is conscious, the nasal passages and mouth may be irrigated with water.

If ingested, do not induce vomiting. If conscious, administer 6 ounces milk (about 175 milliliters) to an adult, or 4 ounces (about 100 milliliters) to a child. Seek medical attention immediately.

10.5 Safety Controls

The gas air conditioning system employs several safety controls which are designed to shut down the unit in the event that a malfunction has occurred. These controls prevent possible damage to the unit, as well as preventing a condition which may be a safety hazard.

The condenser fan assembly is equipped with a sail switch that is designed to sense air flow produced by the fan. This switch has normally open contacts which close when the condenser fan is operating and the air flow is normal. The contacts of the switch are connected in series with the gas valve circuit so that the burner will be inhibited from operating in the event of condenser air flow failure.

The system employs a high-temperature cutout switch to lock out operation should the unit overheat. This may occur under certain malfunctions and unless the system is automatically turned off, damage could occur. The high-temperature switch must be reset each time it is actuated, but the cause of the problem should be investigated and corrected, since repeated resetting of the switch can result in component damage.

The chiller cold water circulating system is designed to operate with water which has been cooled to not less than 40 degrees F (4.4 degrees C). If the water temperature goes below this level for any reason, a cold water cutoff switch will prevent the cooling system from operating until the water temperature increases above 40 degrees F.

The gas control system is monitored and gas flow automatically shut off in the event that ignition does not occur. Systems which employ a pilot light have a thermocouple which senses the heat generated by the pilot. Should it be extinguished, the gas valve is prevented from operating.

Some systems may be protected by a water flow switch that monitors the circulation of water between the chiller and air handler. If water flow is absent, the flow switch cuts off power to the gas valve and prevents burner operation.

10.6 Preliminary Checkout

Certain preliminary checks should be made to determine if there is some obvious fault which is preventing proper system operation. These checks should be made before attempting to operate the unit. The sequence of the checkout procedure as described here is important, and any fault discovered should be corrected before proceeding.

The chiller should be checked for any obvious ammonia odor. Such an odor indicates a leak in the sealed refrigeration system, which must be corrected. Any yellow stain may indicate the source of the leak. Test kits that indicate the presence of ammonia are available.

The orientation of the chiller cabinet should be level for proper operation of the system. Adjustments should be performed if necessary.

The unit should be checked for any obvious obstruction to the condenser air flow, such as accumulated debris or shrubbery. Any interference with proper air flow must be corrected.

The chiller water level should be checked to ascertain that it is correct. A fill test hose is supplied with the unit to allow checking the water level. The system should not be operated with a low water level, and any water leaks must be corrected. If water must be added to the system, sufficient permanent antifreeze must be added to the system to prevent freeze-up at the lowest expected outside ambient temperature, and to preclude the formation of ice during normal operation. Table 10.1 illustrates the freezing point of the coolant for various concentrations of ethylene glycol.

Additionally, a defoaming agent and fungus inhibitor should be added to the system. Antifreeze protection level can be determined by using a hydrometer (available at automobile parts supply outlets) to measure the specific gravity of the water coolant mixture.

The condition, alignment, and tension of all drive belts used in the system should be checked and corrected if necessary. This includes the hydraulic pump, air handler motor, and any other belt that may be used in the system. Motors should be lubricated in accordance with the manufacturer's specifications.

Table 10.1   Chart Illustrating Antifreeze Protection as a
Function of Percentage of Ethylene Glycol

% by Volume Ethylene Glycol	Coolant Freezing Point
10	26 F -3 C
20	18 F -8 C
30	8 F -13 C
35	0 F -18 C

The fluid level of the hydraulic pump should be checked by removing the cap and checking the distance from the top of the pump to the level of fluid. A distance of 1 1/4 inch (3.2 cm) is satisfactory. If the pump is low on fluid, check for leaks, especially at the solution pump. It is important that no hydraulic fluid enters the sealed system where it can cause damage. A hydraulic pump that is very low on oil and has an ammonia odor indicates a defective solution pump diaphragm. If oil must be added to the hydraulic pump, use only approved hydraulic fluid as specified by the manufacturer.

The chiller water circulation system may be checked without burner operation by disconnecting the power leads to the condenser fan motor and disabling the ignition circuit.

With the chiller tank about half full, purge any possible air from the circulating water circuit. This may be done at the water pump itself. Apply power to the system and set the thermostat below room ambient temperature so that the unit calls for cooling. Allow time for a delay circuit which prevents system operation when the unit is first activated. If the water level in the water tank does not go down immediately after the pump starts operation, shut the power off and determine the reason why the water is not being circulated. The pump must not be operated dry; to do so will cause damage. Purge air out of the system if necessary.

The rate of water flow is indicated by the height of the water column at the tank. Refer to the water flow chart supplied by the manufacturer of the system to determine if the proper flow of water is indicated. If it is not within specification, the water flow valve must be adjusted.

The condenser fan height may be checked against the manufacturer's specifications. This adjustment is critical, since proper air flow must pass through the chiller to ensure proper performance. A condenser fan that is out of adjustment may fail to operate the sail switch, thus inhibiting system operation.

The regulated gas pressure may be checked with the unit operating by connecting a manometer to the manifold gas tap at the gas manifold. The correct pressure can be determined by referring to the service manual, and it depends upon the composition of the gas that is feeding the unit. Natural gas will require regulated pressure readings in the area of 3.5 inches of water column. Propane gas requires a 10-inch pressure reading. The gas valve has the provision to adjust pressure.

All air filters in the system should be checked and replaced or cleaned as necessary. Check all air registers and return ducts, and ascertain that they are open. The unit may be set for cooling and checked for proper operation.

Set the thermostat at least 5 degrees below room temperature and allow sufficient time for the system to stabilize (at least 15 minutes). The temperature differential across the air handler cooling coils should be measured, and a change of 18 to 20 degrees F (about 10 degrees C) should be indicated.   If the differential is too great, increase air flow by increasing blower motor speed or resetting the belt on a multisection pulley to the next step. If the air temperature differential across the air handler is too low, the water differential temperature should be checked. If it is within the normal range of 8 to 10 degrees F (about 5 degrees C), the blower may be slowed down to obtain the correct air temperature differential.

The chiller may have thermometer wells which permit accurate temperature measurements of the inlet and outlet water temperatures. The differential should be between 8 and 10 degrees F (about 5 degrees C) for proper system operation. The water should not be at a temperature less than 40 degrees F (4.4 degrees C). If the water differential is not within specifications, the cooling system may be at fault and must be serviced.

Servicing must be performed in the proper sequence. The electrical system is checked first, followed by the gas system, and finally the sealed refrigeration system.

10.7 Electrical System

The electrical system of the unit consists of the air handler blower motor control, thermostat circuit, water circulation pump, condenser fan motor, and hydraulic pump. All of these components must be in proper operating order before any attempt is made to check the operation of the gas or sealed refrigeration system.

When the air conditioning system is first turned on and the thermostat set for cooling, the air handler should operate immediately. The chiller water pump should operate after a time delay of about 45 seconds.

If the system does not operate as indicated, check all circuit breakers and fuses to be sure that power is applied to the system. The low-voltage control transformers, used in the thermostat circuit, should be checked for 24 volts AC output. Any transformer which does not deliver such voltage should be checked for power at its primary terminals, and any possible open windings. The thermostat circuit should be checked for 24 volts AC power and closure of its cooling contacts.

Check all motor control relays for closure. These are operated by 24 volts AC applied to the coil. Measure the voltage at the relay contacts to be sure that it is present. If any relay fails to operate due to absence of power to the coil, check the unit control wiring for loose or bad connections.

The power input to any motor which does not run should be measured. If the readings are within 10% of the rated value, check any capacitor which is used in the motor circuit. Check the motor shaft for excessive bearing friction. A burned-out motor winding can be detected with an ohmmeter by a resistance measurement of the windings.

Line voltage to the chiller should be checked to be certain that it is within 10% of the rated value. The high-temperature switch should be checked to see if it has been actuated. If so, continued resetting of this switch may cause damage to the unit. The cause of the high-temperature lockout must be investigated and corrected. Check the condenser coil assembly for accumulated debris which will inhibit air flow. Clean as necessary.

The system employs a time delay circuit which delays chiller operation about 45 seconds upon turn-on. This and any other time delay circuit in the system should be checked and replaced if defective.

The system control module may be at fault. If all components check out, replacement of the system control may restore proper electrical operation of the system.

10.8 Gas System

If the electrical system is operating properly but the air conditioner is not providing sufficient water cooling, as indicated by a measurement of less than 8 to 10 degrees F (about 5 degrees C) temperature differential between the inlet and outlet chilled water pipes, the gas system must be checked.

If the gas valve fails to open, check the voltage applied to the valve when the thermostat is set for cooling. It should be about 24 volts AC. If no voltage is measured, check the 24 volts supply which feeds the gas valve. Ascertain that the sail switch is actuated by air flow from the condenser fan. Reposition the switch or condenser fan blade if necessary. The contacts of the sail switch are connected in series with the gas valve electrical circuit to prevent burner operation in the absence of condenser air flow.

Check the low-temperature water switch, and flow switch for closed contacts. Either of these components can prevent gas valve operation. The system control module may be defective if the gas valve is not actuated. Replace the module if necessary.

The regulated gas pressure may be checked with a manometer. A fitting at the valve may be installed to allow the manometer to be connected. The correct pressure will be specified by the manufacturer in accordance with the type and composition of the gas, and can be found in the service or installation manual. Natural gas will require a regulated pressure of between about 2 and 4 inches of water column; propane gas requires a gas pressure of 10 inches of water column. Adjustment of gas pressure is made at the gas valve.

It is important to refer to the gas pressure chart which is supplied by the manufacturer of the system. This chart will indicate the correct gas pressure level, which is a function of the system cooling capacity, as well as the composition of the gas. The abbreviated chart illustrated in Table 10.2 is only an example, and should not be used in place of the correct data supplied by the manufacturer of the cooling system.

The gas burner may be inspected for cleanliness and cleaned if necessary. Use a wooden toothpick to clean out the gas orifices. A fiber bristle brush may be used to clean accumulated dirt from the burner ports. When doing so, hold the burner so that the ports face downward to allow the debris to fall away. The gas chamber should be cleaned if necessary.

Pilot-operated gas valves employ a thermocouple to monitor the pilot flame and prevent gas valve operation in the event that the pilot becomes extinguished. A thermocouple consists of two dissimilar metals which are joined at one end and produce an electrical voltage when heated. A

Table 10.2 Typical Chart Illustrating Gas Pressure Adjustment
in Inches of Water Column for Various
Compositions of Natural Gas

BTU/Cu Ft	Specific Gravity			
	0.55	0.60	0.65	0.70
950	3.2 to 3.3	3.5 to 3.7	3.8 to 3.9	4.1 to 4.2
1000	2.9 to 3.0	3.2 to 3.3	3.5	3.8
1050	2.7	2.9 to 3.0	3.2	3.4 to 3.5
1100	2.4 to 2.0	2.6 to 2.7	2.9	3.1 to 3.2

thermocouple may be checked by measuring the open circuit voltage (Figure 10.2) as the pilot flame is held on by manually holding the pilot pushbutton control down. A thermocouple in good condition will exhibit a voltage reading of at least 25 millivolts DC. Any unit which cannot provide this voltage must be replaced.

Some gas burners do not employ a pilot flame; these use solid state electronic ignition systems which generate a very high voltage that produces a spark. Such systems should be checked for broken, corroded, or cracked wires. Check all connections to be sure they are clean and tight. If the ignition spark is erratic, check the ignitor wire. Check the spacing and position of the ignition electrodes. Replace as necessary.

A microammeter can be used to measure sensor current as shown in Figure 10.3. When the system is first turned on the sensor current is a low value, and rises to 7 or more microamperes. The sparking sound of the ignitor should be audible. If the ignitor fails to operate properly, check the position of the electrode. Replace the system control if necessary.

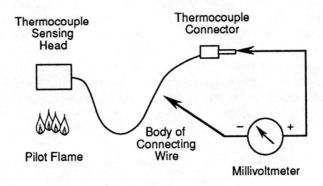

Figure 10.2   Checking Open Circuit Voltage of Thermocouple Using Millivoltmeter

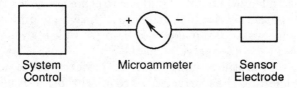

System   Microammeter   Sensor
Control   Electrode

Figure 10.3   Checking Spark Ignition Current
Using Microammeter

10.9 Sealed Refrigeration System

If the electrical and gas systems are both operating properly but the unit does not provide sufficient cooling, the sealed refrigeration section must be checked. A temperature differential of less than 18 degrees F (10 degrees C) between the inlet and outlet air handler temperatures indicates that the refrigeration system may be defective. The chilled water temperature differential should be between 8 and 10 degrees F (about 5 degrees C) in a properly operating chiller.

Prior to investigation of the sealed refrigeration system, the following conditions should be verified as operating properly:

1. Proper air flow through condenser.
2. Chiller cabinet checked for proper orientation (level).
3. Air handler air flow normal.
4. Water circulation between chiller and air handler normal.
5. Thermostat circuit operating.
6. Gas burner OK and operating.
7. Hydraulic pump normal.

The first step in the service procedure is to purge the non-condensable gases. This operation should also be performed if the high-temperature limit switch has been tripped. Purging is done with the unit operating, gas burner on, and at least 5 PSI pressure at the purge valve. A purging hose and bucket of water are required. Personal protection equipment, such as chemical goggles and protective clothing, is recommended.

A hose is connected to the purge valve using adapters as necessary. The open end of the hose should be immersed at least 6 inches (15 cm) below the surface of the water. With the gas burner operating, the purge valve is opened slightly to allow bubbles to come out of the hose. Non-condensable gases will rise to the surface of the water; ammonia will be absorbed by the water as it makes a cracking sound. If a yellow solution appears, close the purge valve immediately.

While the unit is still running, check its performance to determine if the purging operation has corrected the cooling problem. If so, no further investigation is necessary.

After the purge operation is completed, turn the unit off for at least ten minutes to allow the unit to go through its normal time delay shutdown. The bucket of water should be disposed of in an approved, environmentally safe manner.

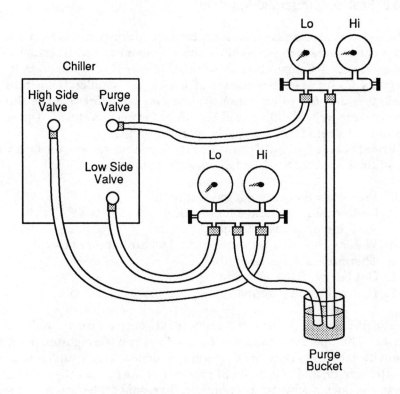

Figure 10.4   Performance Analysis Setup Using Two
Manifold Gauge Sets and Purge Bucket

During the off period connect three gauges to the system as illustrated in Figure 10.4, using a pair of steel manifold gauge sets. The low-pressure gauges are connected to the purge valve and low-pressure access valve. A high-pressure gauge is connected to the high-pressure access valve. The valves on the manifold gauge sets are closed, and the access valves on the chiller unit opened. The open-ended hoses are immersed below the level of the water in the purge bucket.

The ability of the solution pump to move solution is checked by operating the unit with the burner off while observing the purge pressure gauge. Normal indication is a deflection which stops within 5 minutes. A wide deflection that continues for more than 5 minutes indicates a solution pump malfunction, and pump replacement is indicated. No deflection within 5 minutes indicates a possible store-out condition. Further system analysis is necessary.

The gas burner is allowed to operate. The temperature of the solution strainer restrictor should rise, as felt by hand, indicating that weak solution is flowing from the generator to the solution-cooled absorber. The system fault may be diagnosed by observing the three pressure gauges and comparing the results to that indicated in Table 10.3.

10.10 Adding Refrigerant 717

When the air conditioning system is low on ammonia refrigerant, it may be added to the system using the following procedure. Required equipment includes a steel manifold gauge set, cylinder of refrigerant 717, bucket of water, a scale, and safety equipment such as chemical goggles and protective clothing. The refrigerant is added to the low side of the system with the unit in operation. Figure 10.5 shows the charging setup.

The manifold gauge set hoses should be connected as shown. Both valves on the gauge set should be initially closed. Air should be bled out of the hoses by temporarily loosening the hoses at the manifold gauge set and allowing the pressurized ammonia gas, from the cylinder or chiller, to push the air out of the lines. Tighten down the hoses when purged of air.

With the unit in operation and low-side pressure stabilized, open the low-side manifold gauge to allow refrigerant to enter the system. The amount that is added will be indicated by the change in the weight of the refrigerant container. One pound of refrigerant added to the system will

Table 10.3   Diagnostic Chart Showing Possible System Faults

Purge	Low Side	High Side	Diagnosis
No deflection	Does not rise	Does not rise	Refrigerant store-out
No deflection	Falls as high side rises	May not rise but falls if it does	Solution store-out
Steady or erratic deflection	High and steady	Normal to low (yo-yo's)	Partially plugged solution strainer
No deflection	High and steady	Lower than normal. May yo-yo.	Plugged solution strainer restrictor
Increasing deflection beyond 2-4 PSI	Above normal	Rises but falls below normal	Hydraulic or solution pump valve
Normal to stop, erratic operation	Normal and rising pressure, repeating	Falling before reaching normal level, fluctuating	Defective flow control valve
Steady to erratic	Very low	Climbing to very high	Plugged refrigerant restrictor
Over 2-4 PSI deflection	Above normal, may pulsate	Below normal	Cross leak in solution cooled absorber
Normal deflection	Low and steady	Low and steady	Refrigerant charge low
Normal deflection	High and steady	High and steady	Excess refrigerant
Erratic deflection	Below normal	Below normal	Low on solution
Over 2-4 PSI deflection	Higher than normal	Higher than normal	Excess solution

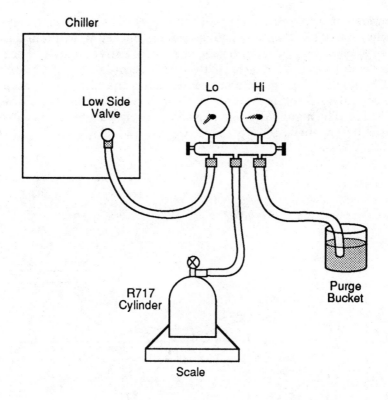

Figure 10.5 Ammonia Gas Charging Setup

increase low-side pressure about 10 PSI. Close the manifold gauge set valve when the desired amount of refrigerant has entered the system.

Allow the system to operate until the system stabilizes. If more refrigerant is needed, repeat the above process. When the low-side pressure has stabilized at the correct level, close the low-side valve on the unit and bleed the refrigerant from the hose by opening the high-pressure valve on the manifold gauge set. Dispose of the water in a safe manner in accordance with environmental regulations.

10.11 Checking Solution Level

Equipment required for checking the level of solution in the system includes two steel manifold gauge sets, a bucket of water, and safety equipment such as chemical goggles and protective clothing. The manifold gauge sets are connected to the unit in accordance with Figure 10.6.

The manifold gauge sets are connected to the high, low, and purge access valves with the center hoses of each set submerged into a bucket of water. The manifold gauge set valves are closed, the access valves on the chiller opened.

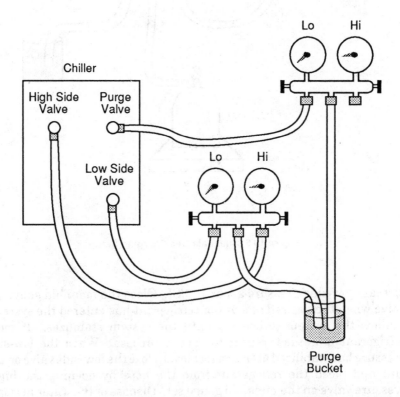

Figure 10.6   Checking Solution Level

The system is placed into operation, and a 95 degrees F (35 degrees C) ambient temperature simulated by partially blocking the condenser air flow so that the high-side pressure is equal to that indicated on a pressure/temperature chart supplied by the manufacturer of the unit. This pressure will be about 300 PSI on a 3-ton unit, and must be sustained for at least 5 minutes.

Hold the loose end of the hose connected to the center of the high- and low-side manifold about 2 inches above the water level. Barely crack open the high-side manifold valve while observing the open end of the hose.

If the level of solution is correct, or low, there will be an initial spurt of solution and then continuous vapor. If previous symptoms in the troubleshooting procedure have indicated a low solution level, add solution to the system as described below. A continuous flow of solution indicates excessive solution level in the system.

10.12 Adding Solution

Equipment required to add solution to the system includes a steel manifold gauge set, four charging hoses, scale, bucket of water, cylinder of solution, cylinder of ammonia, and safety equipment such as chemical goggles and protective clothing. Use the setup of Figure 10.7.

Initially, all valves should be closed. To bleed air from the low-side hose, open both the low-side access valve on the chiller and the low-side manifold valve. Slightly crack open the high-side manifold gauge valve to allow ammonia to flow into the bucket. When it appears, close both manifold gauge set valves.

To bleed air from the solution cylinder hoses, open the vapor valve on the ammonia cylinder. Open the vapor valve on the solution cylinder, then open the liquid valve on the solution cylinder. Slightly crack open the high-side valve on the manifold gauge set. When solution appears in the water bucket close the high-side manifold valve.

Solution is added with the unit in operation and stabilized. The ammonia cylinder and solution cylinder valves are opened. To add solution, open the low-side valve on the manifold gauge set. Do not add more than 2 pounds (0.9 kg) of solution at a time, as indicated by the scale. It is best to add less than is estimated as necessary, then checking the solution level to determine if more is required. Close the low-side manifold gauge valve when the desired amount of solution has been added. Close all other valves.

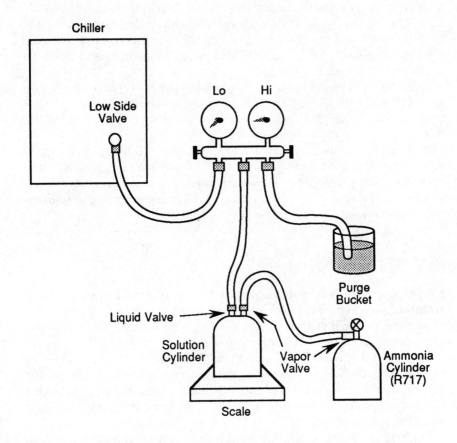

Figure 10.7   Adding Solution to System

To bleed solution from the hoses, slightly open the high-side valve on the manifold gauge set until the solution stops flowing into the bucket. Place a wet rag around the hose fitting on the ammonia cylinder and open the fitting to allow the escaping ammonia to be absorbed by the water in the rag.

Drain any remaining solution from the hoses into the bucket. Dispose of the water in a safe manner in accordance with environmental regulations.

10.13 Removing Solution

Equipment required to remove solution from the unit includes a steel manifold gauge set, scale, purging hose, bucket of water, and safety equipment such as chemical goggles and protective clothing. The setup is illustrated in Figure 10.8.

After turning the system off, allow it to go through its normal 10-minute time delay rest period. Connect the manifold gauge set to the high and low access valves to measure system pressure. Submerge the open end of the purging hose into the water. Record the weight of the bucket of water.

With the unit off, open the purge valve and allow 2 pounds (0.9 kg) of solution to flow into the bucket. Close the purge valve.

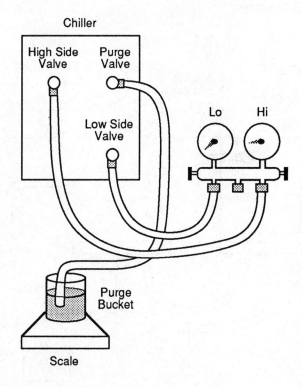

Figure 10.8   Removing Solution from System

Start the unit and allow time for the pressures to stabilize. Check the solution level and, if still too high, repeat the above procedure.

Dispose of the water in a safe manner in accordance with environmental regulations.

10.14 Store-out

Store-out is a condition where the solution is absent from the solution pump, and no liquid is transferred during the first 5 minutes of operation. This symptom is indicated when the pressure gauge connected to the purge valve indicates no deflection within 5 minutes of operation.

Equipment needed to correct store-out is a cylinder of refrigerant 717, two steel manifold pressure gauge sets, and a bucket of water. Figure 10.9 shows the setup.

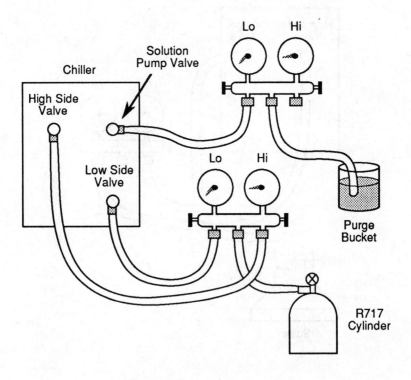

Figure 10.9   Correcting Store-out

Step 1

a. Connect one manifold gauge set to the high and low sides of the system. The low-pressure gauge of another manifold gauge set is connected to the solution pump valve. The center hose of this gauge set is placed in the bucket of water.

b. With the unit not operating, open the access valves to read pressure in the unit. If there is less than 5 PSI indicated, the unit must be charged with refrigerant. Proceed to Step 2.

c. With the gas system disabled, operate the unit electrically for 5 minutes. Open wide the purge valve on the solution pump to purge hard the solution from the solution pump valve. Stop purging if there is any deflection of the gauge needle or appearance of water or liquid, indicating solution moving into solution pump.

d. When gauge needle stops deflection, operate gas burner and observe system pressures. If they rise normally, allow the unit to run about 20 minutes to stabilize, and check operating pressures with pressure/temperature chart supplied by the manufacturer. If the pressure does not rise, turn gas off and proceed to Step 2.

Step 2

a. With the gas disabled (burner off) and the unit operating electrically, open the low-side valve on the manifold gauge set and open the refrigerant cylinder vapor valve. The system access valve should be closed. Read the cylinder pressure. If it is below 100 PSI, place the cylinder in a container of hot water to raise its pressure to 100 PSI minimum.

b. Open the low-side valve on the unit and allow refrigerant to enter until the pressure in the unit has equalized with the refrigerant cylinder pressure.

c. Open the purge valve on the solution pump and purge hard. The appearance of solution or a deflection of the gauge needle indicates solution movement. Stop purging.

d. When needle deflection stops, close the valve on the refrigerant cylinder. Turn the gas burner on.

e. Allow the unit to operate until pressures have stabilized, and compare the readings with the pressure/temperature chart supplied by the manufacturer. If necessary, adjust the refrigerant charge.

f. If there is no indication of solution moving when the unit is purged hard for 5 minutes, close the purge and low-side valve. Check the pressure in the cylinder to be sure it is at least 100 PSI.

Step 3

a. Open the high-side valve on the manifold pressure gauge and allow the pressure to equalize with the cylinder pressure. Repeat Steps c, d, and e of Step 2.

b. When the system is stabilized, check pressures against the manufacturer's specifications.

10.15 Pressure/Temperature Table

The operating pressures of the gas air conditioning system are a function of the ambient temperature of the condenser air flow, as well as the size of the system in BTU cooling capacity. As might be expected, the pressures will rise as air temperature increases. A table of pressure/temperature relationships is an invaluable service aid; with it, the service technician can determine if the sealed refrigeration system is malfunctioning in some way.

The manufacturer of the equipment supplies the proper pressure/temperature table that pertains specifically to each of the units that are produced. Since such pressure readings will not be the same for all units, the chart pertaining to the unit under test must be used. The chart illustrated in Table 10.4 is shown here only as an illustration of a typical pressure/temperature chart.

**Table 10.4   Typical Pressure Temperature Chart
for Ammonia Absorption System**

Ambient Air Temperature	Low-side Pressure (PSI)	High-side Pressure(PSI)
60 F 16 C	20	195
65 F 18 C	25	210
70 F 21 C	29	225
75 F 24 C	33	240
80 F 27 C	38	255
85 F 29 C	43	270
90 F 32 C	48	285
95 F 35 C	53	300
100 F 38 C	57	320
105 F 41 C	61	335
110 F 43 C	65	350
115 F 46 C	68	365

Pressure/Temperature Chart

Pressure/Temperature Chart (PSI versus degrees F and C)

Note: Figures in shaded areas are vacuum specified in inches of mercury.

Note: Shaded Area = Inches of Mercury

TEMP		REFRIGERANTS					
°F	°C	11	12	22	500	502	717
-50	-46	28.9	15.4	6.2	12.8	.0	14.3
-48	-44	28.8	14.6	4.8	11.9	.8	13.3
-46	-43	28.7	13.8	3.5	10.9	1.6	12.2
-44	-42	28.6	12.9	2.0	9.8	2.5	11.1
-42	-41	28.5	11.9	0.5	8.8	3.4	10.0
-40	-40	28.4	11.0	0.5	7.6	4.3	8.7
-38	-39	28.3	10.0	1.3	6.4	5.2	7.4
-36	-38	28.2	8.9	2.2	5.2	6.2	6.1
-34	-37	28.1	7.8	2.0	3.9	7.2	4.7
-32	-36	27.9	6.7	3.9	2.6	8.3	3.2
-30	-34	27.8	5.5	4.9	1.2	9.4	1.6
-28	-33	27.7	4.3	5.9	0.1	10.5	0.0
-26	-32	27.5	3.0	6.9	0.9	11.7	0.8
-24	-31	27.4	1.6	7.9	1.6	13.0	1.7
-22	-30	27.2	0.3	9.0	2.4	14.2	2.6
-20	-29	27.0	0.6	10.2	3.2	15.5	3.6
-18	-28	26.8	1.3	11.3	4.1	16.9	4.6
-16	-27	26.6	2.1	12.5	5.0	18.3	5.6
-14	-26	26.4	2.8	13.8	5.9	19.7	6.7
-12	-24	26.2	3.7	15.1	6.8	21.2	7.9
-10	-23	26.0	4.5	16.5	7.8	22.8	9.0
-8	-22	25.8	5.4	17.9	8.8	24.4	10.3
-6	-21	25.5	6.3	19.3	9.9	26.0	11.6
-4	-20	25.3	7.2	20.8	11.0	27.7	12.9
-2	-19	25.0	8.2	22.4	12.1	29.4	14.3
0	-18	24.7	9.2	24.0	13.3	31.2	15.7
2	-17	24.4	10.2	25.6	14.5	33.1	17.2
4	-16	24.1	11.2	27.3	15.7	35.0	18.8
6	-14	23.8	12.3	29.1	17.0	37.0	20.4
8	-13	23.4	13.5	30.9	18.4	39.0	22.1
10	-12	23.1	14.6	32.8	19.7	41.1	23.8
12	-11	22.7	15.8	34.7	21.2	43.2	25.6

TEMP		REFRIGERANTS					
°F	°C	11	12	22	500	502	717
14	-10	22.3	17.1	36.7	22.6	45.5	27.5
16	-9	21.9	18.4	38.7	24.1	47.7	29.4
18	-8	21.5	19.7	40.9	25.7	50.1	31.4
20	-7	21.1	21.0	43.0	27.3	52.5	33.5
22	-6	20.6	22.4	45.3	28.9	54.9	35.7
24	-4	20.1	23.9	47.6	30.6	57.4	37.9
26	-3	19.7	25.4	50.0	32.4	60.0	40.2
28	-2	19.1	26.9	52.4	34.2	62.7	42.6
30	-1	18.6	28.5	54.9	36.0	65.4	45.0
32	0	18.1	30.1	57.5	37.9	68.2	47.6
34	1	17.5	31.7	60.1	39.9	71.1	50.2
36	2	16.9	33.4	62.8	41.9	74.1	52.9
38	3	16.3	35.2	65.6	43.9	77.1	55.7
40	4	15.6	37.0	68.5	46.1	80.2	58.6
42	6	15.0	38.8	71.5	48.2	83.4	61.6
44	7	14.3	40.7	74.5	50.5	86.6	64.7
46	8	13.6	42.7	77.6	52.8	90.0	67.9
48	9	12.8	44.7	80.8	55.1	93.4	71.1
50	10	12.0	46.7	84.0	57.6	96.9	74.5
52	11	11.2	48.8	87.4	60.1	100.5	78.0
54	12	10.4	51.0	90.8	62.6	104.1	81.5
56	13	9.6	53.2	94.3	65.2	107.9	85.2
58	14	8.7	55.4	97.9	67.9	111.7	89.0
60	16	7.8	57.7	101.6	70.6	115.6	92.9
62	17	6.8	60.1	105.4	73.5	119.6	96.9
64	18	5.9	62.5	109.3	76.3	123.7	101.0
66	19	4.9	65.0	113.2	79.3	127.9	105.3
68	20	3.8	67.6	117.3	82.3	132.2	109.6
70	21	2.8	70.2	121.4	85.4	136.6	114.1
72	22	1.6	72.9	125.7	88.6	141.1	118.7
74	23	0.5	75.6	130.0	91.8	145.6	123.4
76	24	0.3	78.4	134.5	95.1	150.3	128.3

TEMP		REFRIGERANTS					
°F	°C	11	12	22	500	502	717
78	26	0.9	81.3	139.0	98.5	155.1	133.2
80	27	1.5	84.2	143.6	102.0	159.9	138.3
82	28	2.2	87.2	148.4	105.6	164.9	143.6
84	29	2.8	90.2	153.2	109.2	170.0	149.0
86	30	3.5	93.3	158.2	112.9	175.1	154.5
88	31	4.2	96.5	163.2	116.7	180.4	160.1
90	32	4.9	99.8	168.4	120.6	185.8	165.9
92	33	5.6	103.1	173.7	124.5	191.3	171.9
94	34	6.3	106.5	179.1	128.6	196.9	178.0
96	36	7.1	110.0	184.6	132.7	202.6	184.2
98	37	7.9	113.5	190.2	136.9	208.4	190.6
100	38	8.8	117.2	195.9	141.2	214.4	197.2
102	39	9.6	120.9	201.8	145.6	220.4	203.9
104	40	10.5	124.6	207.7	150.1	226.6	210.7
106	41	11.3	128.5	213.8	154.7	232.9	217.8
108	42	12.3	132.4	220.0	159.4	239.3	225.0
110	43	13.2	136.4	226.4	164.1	245.8	232.3
112	44	14.2	140.5	232.8	169.0	252.4	239.8
114	46	15.1	144.7	239.4	173.9	259.2	247.5
116	47	16.1	148.9	246.1	179.0	266.1	255.4
118	48	17.2	153.2	252.9	184.2	273.1	263.5
120	49	18.2	157.7	259.9	189.4	280.3	271.7
122	50	19.3	162.2	267.0	194.8	287.5	280.1
124	51	20.5	166.7	274.3	200.2	295.0	288.7
126	52	21.6	171.4	281.6	205.8	302.5	
128	53	22.8	176.2	289.1	211.5	310.2	
130	54	24.0	181.0	296.8	217.2	318.0	
132	56	25.2	185.9	304.6	223.1	325.9	
134	57	26.5	191.0	312.5	229.1	334.0	
136	58	27.8	196.1	320.6	235.2	342.3	
138	59	29.1	201.3	328.9	241.4	350.6	
140	60	30.4	206.6	337.3	247.7	359.1	

TEMP		REFRIGERANTS					
°F	°C	11	12	22	500	502	717
142	61	31.8	212.0	345.8	254.2	367.8	
144	62	33.2	217.5	354.5	260.7	376.6	
146	63	34.7	223.1	363.3	267.4	385.6	
148	64	36.2	228.8	372.3	274.2	394.7	
150	66	37.7	234.6	381.5	281.1	403.9	
152	67	39.2	240.5	390.8	288.1	413.4	
154	68	40.8	246.5	400.3	295.3	422.9	
156	69	42.4	252.6	410.0	302.5	432.7	
158	70	44.1	258.8	419.8	309.9	442.5	
160	71	45.8	265.1	429.8	317.4	452.6	
162	72	47.5	271.5	440.0	325.1		
164	73	49.2	278.1	450.4	332.9		
166	74	51.0	284.7	460.9	340.8		
168	76	52.9	291.5	471.7	348.8		
170	77	54.8	298.3	482.6	357.0		
172	78	56.8	305.3	493.7	365.3		
174	79	58.8	312.4	505.0	373.7		
176	80	60.8	319.6	516.5	382.3		
178	81	62.8	326.9	528.2	391.0		
180	82	64.7	334.3	540.1	399.9		
182	83	66.9	341.0	552.2	408.8		
184	84	69.1	349.5	564.5	418.0		
186	86	71.3	357.3	577.1	427.3		
188	87	73.5	365.2	589.9	436.7		
190	88	75.7	373.3	602.9	446.3		
192	89	78.1	381.4	616.1	456.0		
194	90	80.5	389.7	629.6	465.9		
196	91	83.0	398.2	643.4	475.9		
198	92	85.3	406.7	657.4	486.1		
200	93	87.8	415.4	671.7	496.4		

B

Temperature Conversion Chart

Fahrenheit to Celsius and Celsius to Fahrenheit

To convert degrees F to degrees C and vice versa, locate the known temperature in the center column, then read its equivalent in either the C or F column.

The conversion of temperature scales is based on the relationships:

$$F = C * 9/5 + 32$$
$$\text{and}$$
$$C = (F\text{-}32) * 5/9$$

°C	F/C	°F
-73.3	-100	-148
-67.8	-90	-130
-62.2	-80	-112
-56.7	-70	-94
-51.1	-60	-76
-45.6	-50	-58
-40.0	-40	-40
-34.4	-30	-22
-28.9	-20	-4
-23.3	-10	14
-17.8	0	32
-17.2	1	33.8
-16.7	2	35.6
-16.1	3	37.4
-15.6	4	39.2
-15.0	5	41.0
-14.4	6	42.8
-13.9	7	44.6
-13.3	8	46.4
-12.8	9	48.2
-12.2	10	50.0
-11.7	11	51.8
-11.1	12	53.6
-10.6	13	55.4
-10.0	14	57.2
-9.4	15	59.0
-8.9	16	60.8
-8.3	17	62.6
-7.8	18	64.4
-7.2	19	66.2

°C	F/C	°F
-6.7	20	68.0
-6.1	21	69.8
-5.6	22	71.6
-5.0	23	73.4
-4.4	24	75.2
-3.9	25	77.0
-3.3	26	78.8
-2.8	27	80.6
-2.2	28	82.4
-1.7	29	84.2
-1.1	30	86.0
-0.6	31	87.8
0	32	89.6
0.6	33	91.4
1.1	34	93.2
1.7	35	95.0
2.2	36	96.8
2.8	37	98.6
3.3	38	100.4
3.9	39	102.2
4.4	40	104.0
5.0	41	105.8
5.6	42	107.6
6.1	43	109.4
6.7	44	111.2
7.2	45	113.0
7.8	46	114.8
8.3	47	116.6
8.9	48	118.4
9.4	49	120.2

°C	F/C	°F	°C	F/C	°F
10.0	50	122.0	27.2	81	177.8
10.6	51	123.8	27.8	82	179.6
11.1	52	125.6	28.3	83	181.4
11.7	53	127.4	28.9	84	183.2
12.2	54	129.2	29.4	85	185.0
12.8	55	131.0	30.0	86	186.8
13.3	56	132.8	30.6	87	188.6
13.9	57	134.6	31.1	88	190.4
14.4	58	136.4	31.7	89	192.2
15.0	59	138.2	32.2	90	194.0
15.6	60	140.0	32.8	91	195.8
16.1	61	141.8	33.3	92	197.6
16.7	62	143.6	33.9	93	199.4
17.2	63	145.4	34.4	94	201.2
17.8	64	147.2	35.0	95	203.0
18.3	65	149.0	35.6	96	204.8
18.9	66	150.8	36.1	97	206.6
19.4	67	152.6	36.7	98	208.4
20.0	68	154.4	37.2	99	210.2
20.6	69	156.2	37.8	100	212.0
21.1	70	158.0	43.3	110	230.0
21.7	71	159.8	48.9	120	248.0
22.2	72	161.6			

C

English and Metric Conversion Factors

Many units are specified with a multiplier prefix, which represents a multiplying factor 10 raised to a positive or negative power. Prefix multipliers which represent factors greater than 1 are capitalized; those which represent factors less than one are not. The most common multiplier prefixes are:

SYMBOL	PREFIX	MULTIPLIER
M	Mega	1,000,000
K	Kilo	1,000
H	Hecto	100
D	Deka	10
d	deci	0.1
c	centi	0.01
m	milli	0.001
u	micro	0.000001

In the following table of conversion factors, the above prefixes may be substituted for those specified if the indicated multiplying constant is replaced with an appropriate one that is modified by the required factor of 10.

TO CONVERT	MULTIPLY BY
Atmospheres to PSI	14.696
Atmospheres to inches of mercury	29.92126
Atmospheres to feet of water	33.94
Atmospheres to Kg/sq cm	1.033
Atmospheres to KPa	101.326
Bars to Kg/sq cm	1.0197
Bars to PSI	14.696
BTU to calories	252
BTU/hr to watts	0.2931
BTU to joules	1054.8
BTU/lb to KJ/Kg	2.326
BTU to calories	252
Centimeters to inches	0.394
Cubic centimeters to cu in	0.061
Cubic inches to cu cm	16.387
Gallons to cubic centimeters	3785
Gallons to cubic inches	231
Grams to ounces	0.0353
HP to watts	746

TO CONVERT	MULTIPLY BY
Inches of mercury to KPa	3.386
Inches of water to mm of mercury	1.86
Inches of water to PSI	0.036
Inches of mercury to PSI	0.491
Inches to centimeters	2.54
Inches to meters	0.0254
Inches to microns	25,400
Kg/sq cm to atmospheres	0.9678
Kg/sq cm to PSI	14.2233
Kilograms to ounces	35.27
Kilograms to pounds	2.205
KPa to inches of mercury	0.2953
Liters to cubic inches	61.025
Liters to gallons	0.2642
Liters to ounces	33.81
Liters to quarts	1.057
Meters to inches	39.37
Meters to yards	1.094
Microns to inches	0.000394

TO CONVERT	MULTIPLY BY
Microns to millimeters	0.001
mm of mercury to inches of mercury	0.039
mm of mercury to PSI	0.019
millimeters to inches	0.039
mils to inches	0.001
Pounds to grams	453.6
PSI to atmospheres	0.068
PSI to inches of mercury	2.036
PSI to inches of water	27.7
PSI to feet of water	2.31
PSI to Kg/sq cm	0.0703
PSI to KPa	6.8948
PSI to millimeters of mercury	51.71
sq. centimeters to sq. inches	0.155
sq. inches to sq. centimeters	6.45
Therms to BTU	100,000
Tons of cooling to BTU/hr	12,000
Torr to mm of mercury	760.0
Watts to BTU/hr	3413

Sources of Air Conditioning Tools, Equipment, and Supplies

Note: Toll-free numbers specified in this appendix were verified at the time of publication.

**Tools, Equipment, and Supplies
(Nationwide locations, coast to coast)**

Johnstone Supply, Inc. (Nationwide)
P.O. Box 3010
Portland, OR 97208

McMaster-Carr Supply Company (Four locations):

Monmouth Junction Road
Dayton, NJ 08810
201-329-3200

also

600 County Line Road
Elmhurst, IL 60126
708-833-0300

also

9630 Norwalk Boulevard
Santa Fe Springs, CA 90670
213-692-5911

also

9100 Fulton Industrial Boulevard
Atlanta, GA 30336
404-346-7000

W. W. Grainger, Inc. (Nationwide)
Distribution Group General Offices
5959 W. Howard Street
Chicago, IL 60648
708-647-8900
Parts toll-free line: 800-323-0620

Parts and Supplies

Brothers Supply (Replacement parts)
34-48 31st Street
Long Island City, NY 11106
800-762-2660

Fasco Industries, Inc. (Replacement motors)
Motor Division
500 Chesterfield Center
St. Louis, MO 63017
314-532-3505

Harry Alter Co. (Parts and supplies)
810 South Vandeventer Avenue
St. Louis, MO 63110
800-356-1027
314-533-0700

Jones & Auerbacher, Inc. (Parts and supplies)
46 Edison Place
Newark, NJ 07102
201-622-2733

Lucas-Milhaupt, Inc. (Brazing supplies)
5656 S. Pennsylvania Avenue
Cudahy, WI 53110
800-558-4503

Parker Hannifin Corp. (Parts and supplies)
15 Depew Avenue
Lyons, NY 14489
315-946-4891

Ranco Controls (Thermostatic and pressure controls)
8115 U.S. Route 42N
Plain City, OH 43064
614-876-8022

Sol Zemel, Inc. (Parts and supplies)
305 Central Avenue
Plainfield, NJ 07061
800-232-3278
201-754-9100

Tesco Distributors, Inc. (Parts and supplies)
300 Nye Avenue
Irvington, NJ 07111
201-399-0333
Fax: 201-399-0599

Tecumseh Products Company (Compressors)
100 E. Patterson Street
Tecumseh, MI 49286
517-423-8411

Tools

Imperial Eastman (Tools and fittings)
An Imperial Clevite Company
6300 W. Howard Street
Chicago, IL 60648
708-967-4769

Snap-on Tools Corporation (Tools and instruments)
8049-28 Avenue
Kenosha, WI 53141
414-656-5200

Test Equipment

Robinair Division (Tools and instruments)
SPX Corporation
Robinair Way
Montpelier, OH 43543
800-822-5561
419-485-5561

TIF Instruments (Test equipment and instruments)
9101 NW 7th Avenue
Miami, FL 33150
305-757-8811

Victor Equipment Company (Brazing equipment, leak detectors)
Airport Road
Denton, TX 76202
817-566-2000

Professional Trade Organizations

Air Conditioning Contractors of America (ACCA)
1513 16th Street N.W., Washington, DC 20036

Air Conditioning and Refrigeration Institute (ARI)
1501 Wilson Boulevard, 6th fl., Arlington, VA 22209

Air Conditioning and Refrigeration Wholesalers
6360 N.W. 5th Way, #202, Ft. Lauderdale, FL 33309

American National Standards Institute (ANSI)
1430 Broadway, New York, NY 10018

American Society of Heating, Air Conditioning, and
Refrigeration Engineers Inc. (ASHRAE)
1791 Tullie Circle, N.E., Atlanta, GA 30329

Associated Builders and Contractors, Inc.
729 15th Street N.W., Washington, DC 20005

Gas Appliances Manufacturers Association
1901 N. Moore Street, #1100, Arlington, VA 22209

International Institute of Ammonia Refrigeration
111 E. Wacker Drive, #600, Chicago, IL 60601

Mechanical Contractors Association of America, Inc.
1385 Piccard Drive, Rockville, MD 20850

National Ass'n. of Plumbing, Heating, Cooling Contractors
P.O. Box 6808, Falls Church, VA 22046

Northamerican Heating and Air Conditioning Wholesalers Ass'n.
1389 Dublin Road, Columbus, OH 43216

Refrigeration Engineers and Technicians Association
111 E. Wacker Drive, #600, Chicago, IL 60601

Refrigeration Service Engineers Society
1666 Rand Road, Des Plaines, IL 60016

S.M. & Air Conditioning Contractors National Association
P.O. Box 70, Merrifield, VA 22116

Bibliography

Modern Refrigeration and Air Conditioning by Andrew D. Althouse, Carl H. Turnquist, and Alfred F. Bracciano, published 1982 by The Goodheart-Willcox Company, Inc., South Holland, Illinois

Handbook of Heating, Ventilating, and Air Conditioning by John Porges and Fred Porges, sixth edition, published 1971 by The Butterworth Group, London, England

Standard Refrigeration and Air Conditioning Questions and Answers, second edition by Steve Elonka and Quaid W. Minich, published 1973 by McGraw-Hill Book Company, New York

Refrigeration and Air Conditioning Technology by Rex Miller, published 1983 by Bennett Publishing Co., Peoria, Illinois 61615

Electrotechnology Volume 4, by Norman W. Lord, Robert P. Ouellette, and Paul N. Cheremisinoff, published 1980 by Ann Arbor Science Publishers, Inc., Ann Arbor, Michigan 48106

Servel Service Manual, Unitary Gas Fired Air Conditioners, published 1988 by The Dometic Corporation, P.O. Box 3792, Evansville, Indiana 47736

Tecumseh Products Company Service Data, revised 1987, published by Tecumseh Products Company, 100 E. Patterson Street, Tecumseh, Michigan 49286

Cadillac Service Information Manual, published 1989 by Cadillac Division, General Motors Corporation, Detroit, Michigan

York Fundamental of Heat Pump Manual and Heat Pump Training Manual, published by York International, P.O. Box 1592, York, Pennsylvania 17405-1592

Ranco "O" and "P" Control Service Manual, published 1987 by Ranco Controls, 8115 U.S. Route 42N, Plain City, Ohio 43064

The Brazing Book, published 1988 by Handy & Harman, 850 Third Avenue, New York, New York 10022

Glossary

Absolute pressure — Pressure, abbreviated PSIA, which is referenced to a perfect vacuum. It is 14.696 PSI less than gauge pressure, PSIG.

Absolute temperature — Temperature measurement referenced to absolute zero, –460 degrees F or –273 degrees C.

Absolute zero — That temperature at which all molecular motion ceases, and at which a substance would be wholly deprived of heat.

Absorbent — A substance which has the capability to take up or assimilate another.

Absorption cycle — A system of refrigeration which uses a refrigerant (usually ammonia), and an absorbent (usually water), to provide heat transfer from a cooler environment to a warmer one.

Accumulator — Component in an air conditioning system which stores liquid refrigerant that did not vaporize in the evaporator.

A coil — The evaporator coil, shaped like the letter "A," which is placed in the plenum of warm air furnaces to provide air conditioning capability.

Adiabatic compression — The process of compressing a gas without the addition or removal of heat energy.

Adiabatic expansion — The process of expansion of a gas without the addition or removal of heat energy.

Air coil — An assembly of tubing through which air flows to cause an exchange of heat energy.

Air conditioning — Control of cooling, heating, and relative humidity of an enclosed area.

Air handler — A system consisting of a blower motor, fan blade, heat exchanger, and plenum to provide a change of temperature of the air passing through.

Air infiltration — The process whereby air passes into or out of an enclosed area through cracks and other openings in the structure.

Air over — A motor design which is cooled by the flow of air generated by the fan blade it drives.

Allen screw — A headless screw with a recessed hexagonal cavity, often used to secure fan blade bushings to motor shafts.

Alternating current — A form of electrical energy, abbreviated AC, which reverses direction at a periodic rate.

Ambient switch — Control which opens or closes a circuit in accordance with the air temperature.

Ambient temperature — The prevailing air temperature.

Ammeter — Electrical instrument that measures current flow in a circuit, and usually calibrated in amperes.

Ammonia — Refrigerant R-717 used in absorption refrigeration systems.

Ampere — Unit of current flow, equal to one coulomb per second.

Anemometer — An instrument which measures air velocity.

ANSI — American National Standards Institute.

Antifreeze — A chemical, such as ethylene glycol, which is added to water to lower its freezing point.

ARI — Air Conditioning and Refrigeration Institute.

ASHRAE — American Society of Heating, Refrigeration, and Air Conditioning Engineers, Inc.

Atmospheric pressure — Absolute air pressure at a given altitude. Under standard conditions at sea level the pressure is 14.696 PSI absolute, or 29.92126 inches of mercury.

Axial — That direction which is parallel to the shaft of rotation.

Azeotropic mixture — A blend of two or more refrigerants which do not combine chemically but produce the desired pressure/temperature characteristics.

Back pressure — The pressure at the suction inlet of a compressor.

Back-seating valve — An access valve which is able to close a refrigerant path at each end of its adjustment travel.

Balance point — The temperature, in a heat pump controlled environment, in which the output of the heat pump in BTU/hr. is exactly equal to the heat loss of the controlled area in BTU/hr.

BAR — A unit of pressure, equal to one atmosphere (14.696 PSI).

Barometer — An instrument which measures absolute atmospheric pressure, which at sea level under standard conditions is 29.92126 inches of mercury.

Barometric pressure — The absolute pressure of the atmosphere. At sea level this is 29.92126 inches (760 mm) of mercury, 14.696 PSIA, 33.94 feet of water, or 101.3 KiloPascals (KPa).

Bimetallic strip — An assembly of two dissimilar metals which, due to uneven expansion characteristics, bends with temperature changes.

Boiling point — The temperature at which a liquid, at a specified pressure, changes to a vapor.

Bowden cable — A flexible cable used to control a mechanism located at a remote location.

Brazing — A method of joining similar or dissimilar metals using a silver-based filler material. It is sometimes referred to as silver soldering.

BTU — British Thermal Unit. One BTU is the amount of heat energy required to raise 1 pound of water 1 degree Fahrenheit.

Bulb — A chamber placed at the end of a capillary tube to store liquid and gas refrigerant, and to provide temperature sensing through expansion and contraction of its contents.

Burnout — The electrical failure of a motor or compressor, resulting in deteriorated wire insulation and/or refrigeration oil.

Calorie — The amount of heat energy required to raise the temperature of 1 gram of water 1 degree Celsius. 1 calorie is equal to 0.004 BTU.

Capacitance — The measure of the charge storing ability of a capacitor, usually specified in microfarads.

Capacitor — Electrical component which, in an alternating current circuit, draws a current which leads the applied voltage by 90 electrical degrees.

Capacitor run motor — An induction motor, sometimes called permanent split capacitor (PSC), which uses a capacitor and start winding that is energized at all times when the motor is running.

Capacitor start motor — An induction motor which uses a capacitor and start winding to provide the initial starting torque. The capacitor is switched out of the circuit when the motor reaches running speed.

Capillary tube — A small-diameter length of tubing which restricts the flow of liquid refrigerant and separates the high-pressure and low-pressure sides of an air conditioning system.

Carbon dioxide — An inert gas which is sometimes used as a purging agent or to pressurize a system for leak testing.

C.C.O.T. — Cycling clutch orifice tube. An automotive air conditioning system which uses an expansion orifice tube for refrigerant control, and depends upon compressor duty cycle for temperature control.

Celsius — A temperature scale, abbreviated C, in which water freezes at 0 degrees and boils at 100 degrees.

Centigrade — A temperature scale, abbreviated C, which is identical to the Celsius scale.

Central air conditioning — A system of air conditioning which cools the entire building, using one centrally located system.

CFC — Chlorofluorocarbon; a family of chemicals which contains halogen elements such as chlorine and fluorine; Freon 12 and 22 are examples of CFCs.

Change of state — The process by which a liquid absorbs latent heat as it boils to a vapor, and a gas gives up latent heat as it condenses to a liquid.

Charge — In refrigeration, the specific amount of refrigerant contained by a sealed system.

Charging — The procedure of placing refrigerant into a system.

Charging station — An apparatus containing a manifold gauge set, vacuum pump, and refrigerant container, which is used to evacuate and charge an air conditioning system.

Check valve — A valve which allows the flow of refrigerant in one direction only.

Chemical instability—The propensity of refrigerant or refrigerant oil to decompose through the action of contaminants and heat in a sealed system.

Chiller — The entire cooling system of an ammonia absorption unit, with the exception of the water-cooled heat exchanger.

Chlorofluorocarbons — CFCs; a family of chemicals, such as Freon, which contains chlorine and fluorine halogen elements.

Clutch — A mechanical device which permits transmission of torque from a driving member to a driven member, and allows disconnection of such driving force.

Coefficient of performance — Abbreviated COP, equal to the ratio of BTU output versus. BTU input as applied to heat pumps.

Cold — The absence of heat.

Comb — A plastic tool used to straighten out distorted fins on evaporator and condenser coils.

Comfort zone — The area on a psychrometric chart in which the ambient temperature and relative humidity are comfortable to humans.

Compound gauge — A pressure gauge which is capable of measuring both pressure and vacuum.

Compression cycle — A system of refrigeration using the mechanical compression of a circulated refrigerant.

Compression ratio—Sometimes called pumping ratio; the ratio of the absolute high-side pressure to the absolute low-side pressure in an air conditioning or heat pump compressor.

Compressor—A mechanical device which pumps a gas from a state of low pressure to high pressure.

Compressor seal — A rotary seal used in open-type compressors which allows the externally driven crankshaft to be brought out of the housing while preventing loss of refrigerant.

Condensate — Water formed as a result of condensation in an air conditioning system.

Condenser — A refrigeration component in which the high pressure, high temperature refrigerant gas gives up its heat and condenses to a liquid.

Condenser fan — A motor-driven fan blade which forces air through an air-cooled condenser.

Condensing unit — That part of a split air conditioning system which contains the compressor, condenser, and condenser fan.

Conduction — Transfer of heat through a medium.

Constrictor — A narrow-diameter tube or small orifice which is used to restrict the flow of refrigerant.

Contaminant — Any solid, liquid, or gaseous component of a sealed system that is not refrigerant or refrigerant oil, such as air, water, dirt, etc.

Continuous cycle absorption system — An ammonia absorption air conditioning system in which the flow of ammonia and solution is continuous as long as the gas burner is in operation.

Convection — The transfer of heat by means of the circulation of a gas or liquid, due to gravitational force.

Cooling coil — The evaporator in an air conditioning system.

Cooling tower — A structure in which water is cooled by evaporation as it is sprayed into the atmosphere.

Coupling — A connecting device which is used to join two refrigerant lines.

Crankcase heater — An electrically operated heating element placed around or inside a hermetic compressor to prevent migration of refrigerant and dilution of lubricating oil.

Current relay — A mechanical device which closes a pair of contacts in response to current flowing through the coil.

Cut-in — The temperature at which a thermostatic switch activates a circuit.

Cut-out — The temperature at which a thermostatic switch deactivates a circuit.

CW, CCW — Clockwise, counterclockwise; terms which describe the shaft rotation of a motor as viewed from the end specified by the manufacturer.

Cycle — A repeating series of events.

Cycling clutch — Operation of the clutch in an automotive air conditioning system which controls the evaporator temperature by means of periodically interrupting compressor operation.

Cylinder — A container which is used to store gas or liquid, usually under pressure.

Defrost cycle — A sequence of events designed to melt accumulated ice on the outside coil of a heat pump system.

Degree-day — A unit which represents a difference of 1 degree between the mean outside temperature and a given standard, such as 65 degrees F in one 24-hour period.

Dehumidifier — A device which removes moisture from the air.

Desiccant — A moisture-absorbing substance which traps and holds residual water in an air conditioning system.

Dew point — The temperature at which the relative humidity in the air becomes 100%, causing it to begin to condense.

Diagnosis — The procedure of analyzing the operation of an air conditioning system.

Differential — The difference between two levels of temperature or pressure.

Direct current — An electrical current that always flows in the same direction.

Discharge — The output port of a compressor, or the process of bleeding refrigerant from a system.

Discharge line — The high-pressure, high-temperature output line of a compressor.

Discharge pressure — The gauge pressure (PSIG) of the discharge gas of a compressor.

Drier — A component in an air conditioning system which contains a desiccant that traps and holds moisture.

Dry bulb thermometer — An ordinary thermometer which indicates the ambient air temperature.

Dry evaporator — An evaporator in an air conditioning system in which all liquid refrigerant is vaporized before returning to the compressor.

Drying agent — The desiccant which is placed in a receiver/drier or accumulator to absorb moisture.

Duty cycle — The percentage of operating time of a device as compared to the total time of one complete on-off cycle.

EER — Energy efficiency ratio; a measurement of air conditioner efficiency calculated by dividing its BTU/hr. rating at 95 degrees F (35 degrees C) outdoor temperature by its power input in watts.

Electronic leak detector — An electronic instrument which measures the change in electrical characteristic of air when contaminated with refrigerant.

End play — The amount of movement, in an axial direction, of the shaft of a motor.

Equalizer line — A tube which connects one part of a component to another, to allow pressures to become equal.

Evacuate — The process of removing all air, gas, moisture, etc., from a sealed system, by means of a vacuum pump.

Evaporation — The process of a liquid changing to a vapor.

Evaporative cooling — A method of cooling air by removing its heat energy through evaporation of water.

Evaporator — The component of an air conditioning system in which the liquid refrigerant absorbs heat as it boils and becomes a gas.

Expansion orifice tube — A liquid refrigerant metering component which contains filter screens that prevent the passage of solid particles.

Expansion valve — An automatic metering device which controls the flow of liquid refrigerant and separates the high-pressure and low-pressure sides of an air conditioning system.

Fahrenheit — A temperature scale, abbreviated F, in which water freezes at 32 degrees and boils at 212 degrees.

Filter/drier — A component in an air conditioning system which contains a fine mesh screen and desiccant to trap small particles and moisture. It may be located in either a liquid or gas line.

Flare — The enlargement at the end of a refrigeration tube which is held leak-tight by a fitting.

Flash gas — The phenomenon of instantaneous vaporization of liquid refrigerant in an evaporator or capillary tube.

Flooded evaporator — An evaporator in an air conditioning system in which not all liquid refrigerant vaporizes to gas before reaching the outlet.

Flush — To purge a system of contaminants, usually with a liquid.

Flux — A heat-activated chemical which absorbs and prevents the formation of oxides during a brazing or soldering operation.

Frame size — A set of physical dimensions for electric motors, established by NEMA.

Freeze protection — The use of an antifreeze compound to prevent the formation of ice in a circulating water system.

Freezing point — The temperature at which a liquid will change to solid when latent heat is removed.

Freon — A trade name for a family of refrigerants.

Frequency — The number of cycles per second (Hertz) of an alternating current power source.

Frost back — A condition by which the suction line of an air conditioning system develops frost or sweating due to liquid refrigerant in the line.

Full load amperes — FLA; the current drawn by a motor when operated at full load.

Gage pressure — The pressure, referenced to atmospheric sea level pressure, as read by a gauge. It is specified as PSIG.

Gas — The vapor state of a substance.

Gauge pressure — A differential pressure, referred to as PSIG, which is the difference between the absolute value of the pressure being measured and atmospheric pressure.

Gauge set — A pair of pressure gauges which are assembled into a manifold to allow the simultaneous monitoring of the high-pressure and low-pressure sides of an air conditioning system, and to evacuate or charge it.

Generator — In ammonia absorption systems, the component in which the strong ammonia solution is heated to produce ammonia vapor.

Halide leak detector — A refrigerant leak detector which uses the property of combustion of a refrigerant to cause a change in flame color to indicate the presence of a leak.

Halide refrigerant — A family of refrigerants, such as Freon, which contains halogen components.

Hard start — A condition by which a compressor may not start under load, which may happen if suction and discharge pressures have not equalized.

Hard start circuit — Extra components added to a compressor circuit to increase starting torque.

HCFC — Hydrochlorofluorocarbon; a family of chemical refrigerants which pose little or no threat to the ozone layer and are being considered as replacements for CFC refrigerants.

Head pressure — The discharge pressure of a compressor.

Heat exchanger — A device which is designed to transfer heat energy between two mediums, such as refrigerant and air or refrigerant and water.

Heat gain — The amount of heat energy which enters or is produced within an enclosed area from heat sources such as solar, electrical power, infiltration, habitation, etc.

Heat load — The amount of heat energy, in BTU/hr., presented to the evaporator heat exchanger in an air conditioning system.

Heat pump — A mechanical refrigeration cycle air conditioning system which contains a reversing valve and is able to transfer heat energy, from an area of lower temperature to one of higher temperature, in either of two directions.

Hermetic compressor — A compressor assembly in which the induction motor and compressor are both sealed in the same shell.

Hertz — Unit of cycles per second in alternating current systems.

HFC — Hydrofluorocarbon; a family of chemical refrigerants which pose little or no threat to the ozone layer and are being considered as replacements for CFC refrigerants.

High-pressure cutout — A pressure control which interrupts compressor operation when discharge pressure exceeds a safe value.

High-pressure line — The refrigerant tubing which carries the high-pressure compressor discharge gas, or liquid refrigerant, from the condenser.

High side — The high-pressure side of an air conditioning system between the discharge port of the compressor, through the condenser, and to the refrigerant metering device.

Humidistat — A control switch that reacts to changes in the level of relative humidity.

Humidity — Relative humidity; the amount of moisture in the air expressed as a percentage of the maximum amount that can be contained at that temperature.

HVAC — An acronym used to identify the heating, ventilating, and air conditioning industry.

Hydraulic pump — A pulsating pump which, by means of pressurizing hydraulic fluid, operates the diaphragm of an ammonia solution pump.

Hydrometer — An instrument which measures the specific gravity of a liquid, such as antifreeze solution.

Hygrometer — An instrument which measures the level of relative humidity.

Induction motor — An AC-operated motor containing stator windings which are excited by an AC power source, and a rotor which carries induced current.

Infiltration — The normal exchange of air to or from an enclosed area through crevices, doors, windows, etc., due to pressure differences between the outside and inside.

Impedance protected — A type of motor design which is able to withstand continuous excitation, under locked rotor conditions, without damage.

Kelvin — An absolute temperature scale, abbreviated K, which uses the same divisions as the Celsius scale and is referenced to absolute zero. Degrees K equals degrees C plus 273.

KiloPascal — KPa, a unit of force per unit area. Atmospheric pressure, 14.696 PSIA, is equal to 101.255 KPa.

Kilowatt — A unit of electrical power equal to 1000 watts.

Latent heat — The heat energy required to change the state (gas to liquid, liquid to gas) of a substance. Latent heat is given up when a gas condenses to a liquid. Latent heat is absorbed by a liquid when vaporizing to a gas.

Leak detector — An instrument which reacts to a refrigerant that is escaping from an opening in a sealed system, and pinpoints the source of leakage.

Limit control — A pressure or temperature control which interrupts system operation when a predetermined limit has been reached.

Liquid line — The refrigerant line containing liquid refrigerant and connected between the condenser and restricting device.

Locked rotor amperes — LRA; the current drawn by a motor when the rotor's shaft is not rotating.

Low-pressure line — The refrigerant line connecting the outlet of the evaporator to the suction inlet of the compressor.

Low side — That part of the air conditioning system between the restricting device, through the evaporator, and to the suction inlet of the compressor.

Manifold gauge set — Low-pressure and high-pressure gauges mounted on an assembly consisting of hand-operated valves and hoses, used for diagnosing, evacuating, and charging an air conditioning system.

Manometer — An instrument which is used to measure low pressures, using a column of liquid such as water or mercury.

Megger — An instrument which measures high values of insulation resistance of an electrical device by impressing 500 volts or more between the electrical terminals of the unit under test and ground or frame.

Megohm — A unit of resistance, equal to 1,000,000 ohms.

Microammeter — An instrument which is capable of measuring current levels as small as one millionth of an ampere.

Microfarad — A unit of capacitance equal to 1/1,000,000 farads.

Micron — A unit of length equal to one millionth of a meter. This unit is often used, in place of inches of mercury, to specify the vacuum obtainable from high-grade vacuum pumps which can develop a vacuum of less than 10 microns (a near-perfect vacuum).

Microprocessor — A solid state integrated circuit component which is composed of transistors, diodes, and resistors, and capable of performing calculations or computer operations in accordance with preprogrammed instructions and data input.

Moisture — Water or humidity which is present in the sealed system.

Moisture indicator — A chemical which changes color in the presence of moisture.

Muffler — A device which is placed at the discharge line of the compressor to minimize pumping noises.

NEMA — National Electrical Manufacturers Association.

Non-condensable gas — A gas, such as air, which will not change to liquid when under the normal operating temperatures and pressures of an air conditioning system.

OEM — Original equipment manufacturer.

Off cycle — That part of the cycle of operation of a system when it is not running.

Ohm — The unit of resistance. A 1-ohm load will draw 1 ampere from a 1-volt power source.

Ohmmeter — An instrument that measures resistance.

Oil bleed line — Capillary tube which allows refrigerant oil that collects in a component to return to the compressor.

Oil charge — A quantity of oil, under pressure or not, which is to be added to an air conditioning system. Also, the amount of oil which is contained in a sealed system.

Operational test — The performance test of an air conditioning system used to determine if a malfunction exists.

Orifice — A fixed opening component which is designed to meter the flow of liquid refrigerant in accordance with the pressure differential across it.

Overload — An electrical component which monitors current and temperature, and opens the circuit when a safe level has been exceeded.

Performance test — The process of taking pressure and temperature readings to determine if an air conditioning system is operating properly.

Phase change — The physical change of a substance from liquid to vapor or vapor to liquid in which latent heat is absorbed or released.

Plenum — An enclosure above the heat exchanger in a warm air furnace.

Plenum blower assembly — The air chamber in which warm humid air forced through the evaporator coils gives up its heat and moisture to the liquid refrigerant.

POA valve — Pressure-operated absolute valve which automatically maintains a specified minimum evaporator pressure to prevent the buildup of frost on the coils.

Pole — Part of the stator of an induction motor. The number of poles is always an even number, and determines the rpm rating of the motor.

Polyphase — A multiphase electrical system, usually two- or three-phase.

Potential relay — An electrical component which opens or closes a set of contacts in response to the voltage impressed upon its coil.

Power element — That part of a thermostatic control which reacts to temperature changes and causes a mechanical motion.

Power factor — The ratio of watts to volt-amperes in an alternating current system, and defined as the cosine of the angle of lead or lag of the current, with respect to the voltage.

PPM — Parts per million; a measure of concentration. 1 PPM is equal to 0.0001%.

Pressure — Force per unit area, measured in pounds per square inch, inches or millimeters of mercury, inches of water, kilograms per square centimeter, etc.

Pressure drop — The loss in pressure from one part of a refrigeration circuit to another.

Pressure line — Discharge line of a compressor.

Pressure regulator — A device which maintains a predetermined pressure level at its output, when input pressure exceeds that level.

Pressure/temperature chart — A compilation of the boiling point temperatures of a refrigerant, listed in accordance with various pressure levels.

Process tube — A short stub of refrigeration tubing, usually located on the compressor, for evacuation and charging the system.

PSI — Pounds per square inch.

PSIA — Pounds per square inch absolute. The pressure of a gas or liquid specified with reference to a perfect vacuum.

PSIG — Pounds per square inch gauge. The pressure of a gas or liquid specified with reference to atmospheric pressure at sea level (14.696 PSIA).

Psychrometer — A wet bulb hygrometer that is used to measure relative humidity levels.

Psychrometric chart — A graph in which relative humidity levels are plotted as a function of the dry bulb and wet bulb temperature.

PTC — Positive temperature characteristic; the property of a resistor or solid state device to exhibit a positive change in resistance value with increasing temperature.

Pump down — The process of evacuating an air conditioning system.

Pumping ratio — The ratio of the absolute head pressure to the absolute suction pressure in an air conditioning or heat pump compressor.

Purge — The process of forcing contaminants out of a component or system by means of a pressurized inert gas or liquid.

Purging bucket — A container of water which is used to absorb ammonia gas or ammonia solution when servicing an absorption system chiller.

R-11 — Trichloro-monofluoro-methane refrigerant.

R-12 — Dichloro-difluoro-methane refrigerant.

R-22 — Monochloro-difluoro-methane refrigerant.

R-40 — Methyl chloride refrigerant.

R-113 — Trichloro-trifluoro-ethane refrigerant.

R-134A — Tetra-fluoro-ethane, an ozone-friendly HFC refrigerant.

R-160 — Ethyl chloride refrigerant.

R-170 — Ethane refrigerant.

R-290 — Propane refrigerant.

R-500 — Azeotropic mixture of R-12 and R-152A.

R-502 — Azeotropic mixture of R-23 and R-13.

R-503 — Azeotropic mixture of R-13 and R-23.

R-504 — Azeotropic mixture of R-32 and R-115.

R-600 — Butane refrigerant and fuel.

R-611 — Methyl formate refrigerant.

R-717 — Ammonia refrigerant.

R-744 — Carbon dioxide refrigerant.

R-764 — Sulphur dioxide refrigerant.

Radial — That direction which is perpendicular to the shaft of rotation.

Rankine — An absolute temperature scale, abbreviated R, using the same divisions as the Fahrenheit scale and referenced to absolute zero. Degrees R equals degrees F plus 460.

Receiver/drier — A component in an expansion valve controlled air conditioning system which stores liquid refrigerant, and contains a desiccant to trap moisture.

Refrigerant — The liquid and gas substance in an air conditioning system which provides heat transfer by its change of state.

Refrigeration — The mechanical system of transferring heat from a location of low temperature to one of higher temperature.

Refrigeration cycle — The process of transferring heat by means of compression and expansion of a refrigerant.

Relative humidity — The amount of moisture in the air, expressed as a percentage of the maximum amount of moisture that can be contained in the air at the same temperature.

Relay — An electromechanical device which controls the opening or closing of its contacts by means of current flow in its coil.

Restrictor — A device which impedes the flow of a gas or liquid to cause a pressure drop.

Reversing valve — A device used in a heat pump that is used to change the direction of flow of refrigerant.

Room temperature — 68 to 72 degrees F, or 20 to 22.2 degrees C.

Room unit — A self-contained air conditioning system designed to be mounted in a window or wall to cool one or more rooms.

Rotor — The rotating portion of an induction motor which carries the induced current.

RPM — Revolutions per minute; in an induction motor equal to approximately (120)(frequency)/(# of poles).

Sail switch — A switch, connected into the control circuit of a cooling system, that responds to the force of air moving through an air-cooled condenser and prevents system operation if the flow of air is insufficient.

Saturated vapor — Vapor, at any given temperature and pressure, that would begin to condense to liquid if its temperature fell or pressure increased.

Schrader valve — A spring loaded valve which permits access to the low and high sides of an air conditioning system. The valve is automatically opened when the manifold gauge set hose is connected.

Screen — A fine mesh filter which is placed at the inlet of expansion valves or filter/driers to prevent the passage of solid particles.

Sealed unit — A hermetic air conditioning system.

SEER — Seasonal energy efficiency ratio; a measurement of heat pump or air conditioner efficiency calculated by dividing its BTU output at 82 degrees F (27.8 degrees C) outside air temperature by its power input in watts.

Sensible heat — Heat energy which causes a temperature change in a substance.

Service factor — A number, more or less than 1, by which the HP rating of a motor is multiplied to determine the maximum safe continuous load rating.

Service valve — An access valve used for diagnosing, evacuating, and charging an air conditioning system.

Shaded pole — An induction motor design which employs an induced current path around the stator, to generate starting torque.

Short cycling — The failed attempt of a compressor to start before the high-side and low-side pressures have equalized.

Sight glass — A window which permits the visual observance of liquid refrigerant as it flows between the condenser and expansion device. It also may contain a moisture-indicating chemical.

Silver brazing — A method of joining two metals by the use of a silver bearing filler material at temperatures above 800 degrees F (427 degrees C) and below the melting point of the materials being joined. Also called silver soldering.

Single phase — An alternating power source which is supplied by two wires.

Sling psychronometer — A humidity measuring instrument containing wet bulb and dry bulb thermometers and swung through the air to obtain a true wet bulb temperature reading.

Slope coil — A flat evaporator coil placed at an angle in a furnace plenum.

Slugging — A condition in a compressor when liquid enters the suction inlet, causing hammering of the moving parts.

Solar heat — Heat energy derived from the sun by means of a suitable heat exchanger.

Soldering — A method of joining two metals using alloys of tin at temperatures of about 450 degrees F (232 degrees C).

Solenoid valve — A valve that is mechanically actuated by means of an electric current passing through a coil.

Solid state — Electronic components which consist of silicon, germanium, or other semiconductor materials, and used in air conditioning control circuitry.

Solution-cooled absorber — SCA; the component in an ammonia absorption system in which the low-pressure ammonia vapor is absorbed by the weak ammonia solution.

Solution pump — The liquid pump which circulates the ammonia solution between the generator and absorber.

Specifications — Data supplied by the manufacturer of an air conditioning system as related to temperature, pressure, power consumption, etc., when operating properly.

Split phase motor — An induction motor containing a start winding which is magnetically displaced from the main winding.

Split system — An air conditioning system in which the compressor/condenser assembly is placed at a remote location from the evaporator assembly, and connected together by means of tubing.

Squirrel cage — A type of fan blade design which moves air in a radial direction.

Standard conditions — A temperature of 68 degrees F (20 degrees C), atmospheric pressure of 29.92 inches of mercury, and 36 percent relative humidity.

Starting relay — A relay which connects a motor-starting capacitor or winding into the circuit only during the starting sequence.

Store-out — A condition in an ammonia absorption system in which the ammonia solution is not present at the solution pump, and does not flow when the unit is first started.

Subcooling — Cooling of a liquid refrigerant below its condensing temperature, in accordance with the pressure.

Suction line — The refrigerant line between the evaporator and inlet port of the compressor.

Suction pressure — That pressure which exists at the inlet port of the compressor.

Suction side — That portion of the air conditioning system located between the refrigerant metering device and the inlet port of the compressor.

Suction throttling valve — An evaporator pressure control valve which automatically maintains sufficient evaporator pressure to prevent frost buildup.

Superheat — The temperature differential between the outlet of the evaporator and the boiling temperature of the liquid refrigerant.

Superheated vapor — Any gas which is at a temperature above the boiling point of the substance at a given pressure.

Synchronous speed — The theoretical maximum rpm of an induction motor equal to (120)(frequency)/# of poles.

System — All the components that make up an air conditioning unit.

Temperature — Heat energy measured by Celsius, Fahrenheit, Kelvin, or Rankine scales.

TENV — Totally enclosed non-ventilated; a type of motor design which does not require internal air cooling.

TE valve — Thermostatic expansion valve which meters liquid refrigerant in accordance with evaporator temperature.

Thermal cutout — A protective device which monitors temperature and opens its set of contacts when a predetermined temperature limit has been reached.

Thermal relay — A relay which is actuated by means of an element that is heated by electrical energy.

Thermistor — A resistor which has a positive or negative coefficient of resistance so that its value changes with temperature.

Thermocouple — A device which, when heated, develops an electrical potential.

Thermometer — An instrument which measures temperature.

Thermostat — A device which reacts to ambient temperature changes and causes a pair of switch contacts to close or open.

Thermostatic expansion valve — Commonly called TEV or TXV. Used to meter liquid refrigerant flow in accordance with evaporator heat load and, thus, its temperature.

Thermostatic switch — A temperature activated control which is used to control the on and off times of a compressor.

Three phase — An alternating current power source which is supplied in 3-wire delta or 4-wire wye configurations.

Ton of refrigeration — Cooling effect produced by melting 1 ton of ice in a 24-hour period, and equal to 12,000 BTU per hour.

Torque — A turning force, usually specified as that required to properly assemble a refrigeration connection. Specified in foot-pounds in English measure.

Transformer — An AC operated component which lowers or raises the supply voltage.

Transistor — A solid state component, usually composed of silicon, which has the ability to amplify current.

TXV — Alternate nomenclature for thermostatic expansion valve.

Universal motor — A motor which will operate from an AC or DC power source.

Vacuum — Any pressure level which is less than 14.7 PSIA at sea level.

Vacuum gauge — An instrument which measures the level of vacuum or negative pressure.

Vacuum pump — Mechanical device which is capable of transferring air and vapor so that an air conditioning system may be evacuated.

Viscosity — A specified property of a liquid in accordance with its resistance to flow. Usually applied to refrigeration oil as used in air conditioning systems.

Volatile liquid — Any substance which evaporates readily at room temperature, such as R-11.

Voltage — The pressure or driving force of an electrical power source.

Volt-amperes — The product of voltage and current in an electrical load.

Voltmeter — An instrument used for the measurement of voltage level.

Watt — The unit of electrical power. 1 watt is dissipated in a load that is driven by 1 volt and carries 1 ampere of current.

Wet bulb thermometer — An ordinary thermometer which has its sensing bulb continuously moistened so that it reacts to the cooling effect of evaporation.

Index